P9-AZV-238

More Praise for *The Most Dangerous Animal*

"Smith has provided a cogent answer to the deeper why question of war; not why Iraq? or why Afghanistan? or why Darfur?, but why war at all? Smith's answer—that war is buried deep in our evolutionary past—will be controversial, but his case is irrefutable. We have seen the enemy in the mirror, and until we gather the courage to accept our true nature, men will fight and people will die. Every politician should read this book before deciding on war."

—Michael Shermer, publisher of *Skeptic* magazine and author of *The Science of Good and Evil: Why People Cheat, Gossip, Care, Share, and Follow the Golden Rule*

"Here is the unvarnished tale of human gangs, driven by built-in survival mechanisims and an uncanny ability for self-deception, romping through history—raiding, pillaging, terrorizing, waging wars, and committing large-scale atrocities in the name of abstract gods, holy lands, master races, and political systems. David Smith's rapid-fire account of our uniquely lethal nature makes a mockery of our dreams for peace. We could always try, though, but seeing ourselves as we truly are is a necessary first step. This book shows us how."

—Anouar Majid, author of *Freedom and Orthodoxy: Islam and Difference in the Post-Andalusian Age* and *Unveiling Traditions: Postcolonial Islam in a Polycentric World*

"This is a brilliant book. It weaves together a wealth of insights from science, history, literature, philosophy and contemporary affairs into an accessible, lucid, and cogently argued defense of the role of human nature in war."

—Robert L. Holmes, professor of philosopy, University of Rochester, and author of *On War and Morality*

"A remarkable and accessible book that provides original and compelling insights into the human capacity for war. Professor Smith's keen psychological analysis reveals how we unconsciously deploy self-deceptive strategies to override our horror at human bloodshed in order to indulge our universal penchant for intergroup violence. A must-read for anyone interested in the psychological depths of human nature."

—Barbara S. Held, Barry N. Wish Professor of Psychology and Social Studies, Bowdoin College, and author of *Psychology's Interpretive Turn: The Search for Truth and Agency in Theoretical and Philosophical Psychology*

THE MOST DANGEROUS ANIMAL

ALSO BY DAVID LIVINGSTONE SMITH

Why We Lie

THE MOST DANGEROUS ANIMAL

Human Nature and the Origins of War

David Livingstone Smith

ST. MARTIN'S PRESS NEW YORK

 For information, address St. Martin's Press, 175 Fifth Avenue, New York, N.Y. 10010.

www.stmartins.com

Book design by Christopher M. Zucker

Library of Congress Cataloging-in-Publication Data

Smith, David Livingstone, 1953–
The most dangerous animal : human nature and the origins of war / David Livingstone Smith.
p. cm.
Includes bibliographical references.
ISBN-13: 978-0-312-34189-3
ISBN-10: 0-312-34189-X
1. War. I. Title.

U21.S617 2007
355.02—dc22

2007015297

First Edition: August 2007

10 9 8 7 6 5 4 3 2 1

To the fallen

CONTENTS

ACKNOWLEDGMENTS

THIS BOOK WOULD NEVER COME about were it not for the efforts of my agent, Michael Psaltis, and Ethan Friedman, my editor at St. Martin's Press, both of whom displayed the patience of Job as I announced one new deadline after another for the delivery of the manuscript. Ethan moved on to greener pastures before the project came to fruition, and my second editor, Daniela Rapp, picked up the baton and skillfully guided the project to completion. Thanks also to Donald J. Davidson, my amazingly erudite copy editor.

I am grateful to Rob Deaner, Carl Rollyson, Astrida Tantillo, Daniel Wilson, Alice Andrews, Michael Williamson, and Tanya Callan for graciously sharing their ideas with me and pointing me toward sources of information, and to Andrew Becevich, David Barash, and Michael Shermer for their support of the project. Every author needs a home, and I would like to thank Jacque Carter, Paul Burlin, and Linda Sartorelli for giving me a base at the University of New England and for supporting my scholarly efforts. Very special thanks are due to Anouar and Melissa Majid, who provided encouragement and erudition, as well as considerable quantities of nourishment for body and soul.

The literature on war is immense, but the vast bulk of it has little to do with human psychology. I am deeply indebted to the veterans, journalists, and psychologists who have ventured into these deep waters and provided important insights without which this book could not have been written. It is difficult to single out the writers who have influenced me most, but the list certainly includes S. L. A. Marshall,

Chris Hedges, Dave Grossman, Barbara Ehrenreich, Richard Wrangham, J. Glen Grey, Richard Holmes, Joanna Bourke, Robert Bigelow, John Dower, Robert O'Connell, Irenäus Eibl-Eibesfelt, and Johan M. G. Van der Dennen.

Finally, no words can express my gratitude that I owe to my wonderful wife, Subrena, for keeping me reasonably sane, for inspiring me to explore new intellectual horizons, and for shouldering many of the practical burdens of life so that I might have time to think and write.

PREFACE

> Man beset by anarchy, banditry, chaos and extinction must at last resort turn to that chamber of horrors, human enlightenment. For he has nowhere else to turn.
>
> —ROBERT ARDREY, *AFRICAN GENESIS*

RIGHT NOW, AS YOU READ THIS, somebody, somewhere, is planning a war. It may be a genocide, an invasion, a revolution, or even the detonation of a nuclear weapon, but whichever it is, you can be certain that it will destroy bodies, wreck lives, and breed misery for generations to come. Almost 200 million human beings, mostly civilians, have died in wars over the last century, and there is no end of slaughter in sight. The threat hangs over all of us, constant and unrelenting.

War can be approached from many angles. We can consider it from the standpoint of economics, politics, history, ideology, ethics, and various other disciplines. All of these are important, but there is one dimension that underpins them all: the bedrock of human nature. To understand war, we must understand ourselves. Now, honest self-inquiry is not a feel-good activity. It asks unnerving questions and tries to answer them in the unforgiving light of truth rather than the soft glow of wishful thinking. There are some truths that no one likes to hear, but it is precisely these that we need to pursue if we are to understand where war lives in human nature.

Historically, there have been two broad, sharply polarized views of the relationship between war and human nature. One is that war is human nature in the raw, stripped of the façade of contrived civility behind which we normally hide. In most recent incarnations of this ancient theory, the taste for killing is said to be written in our genes. The other is that war is nothing but a perversion of an essentially kind, compassionate, and sociable human nature and that it is culture, not biology, which makes us so dangerous to one another. In fact, both of these images are gross oversimplifications: both are true, and both are false. Human beings *are* capable of almost unimaginable violence and cruelty toward one another, and there is reason to believe that this dogged aggressiveness is grounded in our genes. But we are also enormously sociable, cooperative creatures with an elemental horror of shedding human blood, and this, too, seems to be embedded in the core of human nature. Strange as it may sound, I believe that war is caused by *both* of these forces working in tandem; it is a child of ambivalence, a compromise between two opposing sides of human nature.

I come to this task not as a soldier,[1] but as a philosopher—a paradigmatic ivory-tower dweller. I have never set foot in a war zone, and I am far more familiar with the rarefied violence of philosophical debate, of words tearing into arguments, than I am with bullets tearing through living flesh. Some readers may feel that this background undermines my credibility, because I don't really know what war is like. In fact, I do not pretend to *describe* what war is like. For that, it is best to go to books by combat veterans, men and women with firsthand experience of it. My task is a different one: I want to explain how war is rooted in human nature. If you want to know what it is like to have cancer, ask a cancer patient. However, if you want an *explanation* of how cancer works—what it is, what causes it, and how to treat it—you are better off consulting someone who studies it. What is true of cancer is also true of war. It is partially for this reason that I have drawn primarily on the literature—both the theoretical and historical literature and the personal observations found in memoirs and correspondence—rather than on interviews of military personnel returning from the "sharp end" of battle. There are also other reasons. Combat typically takes place in a fog of confusion. In fact, the word "war" is derived from the Old English "wyrre," which means "to bring into confusion." Veterans often describe their memories of combat as disjointed, dreamlike, and obviously distorted. Some experiences are obliterated by traumatic amnesia, and other men's recollections are sometimes recruited to fill in the gaps. However valuable firsthand accounts of the psychology of battle are—and there is no disputing that they *are* extremely valuable—they must also be treated with caution. Soldiers' recollections are not a photographic record of what actually happened; they are conglomerations of fact and fiction, formed under conditions of extreme psychological stress. The problem is nicely exemplified in a remark by Ned Frankel, cited by John Ellis in his wonderful book *The Sharp End.* Some veterans, Frankel writes, are anxious to speak of their experiences in combat:

> But once they start, even the most articulate of them fall tongue-tied. What was Iwo Jima like? It was . . . it was . . . it was fucking rough man! I know that, but what was it like? Really . . . really . . . really tough! So the very experience of war, what would seem to be the prerequisite for describing it, precluded any actual, palpable narrative.[2]

Peculiar things happen to the mind in war, and the person who actually tastes what combat is like is not necessarily well placed to understand the psychological processes underlying them. One reason for this is the fact that war is all about killing other human beings, and as such it violates one of our most profound, most visceral taboos—a taboo that goes to the heart of what it means to be a social primate. Expecting soldiers to be truthful with themselves about the experience of killing, especially killing in close combat, is too much to demand of anyone. All of these factors working together require us to approach the truth somewhat obliquely—looking at the clues that seep out around the edges of experience rather than asking direct questions point-blank and eliciting direct but ultimately uninformative answers.

Of course, in approaching the subject in this way, I open myself to the charge of cherry-picking the evidence that suits me. This is a reasonable concern. Sure, I run the risk of seeing what I want to see and nothing else. But I'm willing to take that risk, because the potential payoffs are so great. In the end, it is my conclusions, rather than the path I took to reach them, that really counts.

I wrote this book for a wide audience because understanding war is too important a matter to be left in the hands of an academic cabal. The readers that I want to reach out to are the men and women who will go to war, their parents and their spouses, returning veterans struggling to make sense of their experiences, and, most of all, the citizens who elect men and women to high public office—men and women who may decide to take their nation to war, for good reasons or for bad ones. With this broad readership in mind, I have tried to keep my language as clear and nontechnical as possible. However, a certain amount of unfamiliar terminology is inevitable in any analysis that goes beyond the stale platitudes of common sense, so I have made a point of explaining all of the scientific and philosophical concepts as they arise in the text. You won't need an extensive background in science or philosophy to engage with this book. All you will need is a certain amount of patience and an open mind.

This is not an antiwar book, in any obvious sense of the word. I believe that war is a tragic fact of life, and that it is sometimes inevitable. But I also believe that it is vitally important to understand just what we are getting into when we decide to go to war. More specifically, we need to understand the irrational *allure* of mass violence, the forms of self-deception that are its handmaidens, and the true human costs concealed behind fantasies of valor and righteousness. It is only by taking these factors into account that we can decide where we stand—as individuals and as nations—with respect to the momentous and potentially catastrophic decisions that confront us when we consider going to war.

Although scientific in tone, I hope that this book will not be limited to an exclusively secular readership. Not everyone agrees that science is helpful for understanding human nature, and explanations of human nature that draw their inspiration from evolutionary biology are, at least in the United States, especially contentious. Sadly, there is a widening gap between the scientific consensus about the evolutionary origins of humankind and the religious beliefs of many ordinary citizens. This book *is* unashamedly rooted in an evolutionary biological perspective, but I earnestly hope that even if you are uneasy with this you will pursue my thesis to its conclusion rather than discarding it in exasperation. Even if you disagree with my general position, you may chance upon some ideas that are worth considering. And who knows? I might even succeed in changing your mind.

In the pages to follow, I will move promiscuously between disciplines—from psychology, to philosophy, to prehistoric archaeology, with forays into anthropology, psychoanalysis, and even microbiology, thus running the risk of being regarded as jack-of-all-trades and master of none. Nature does not respect the artificial boundaries between disciplines carved out by university departments, and anyone interested in the big picture of any natural phenomenon—including the phenomenon of war—must emulate her subversion of traditional borders. No doubt, I will sometimes put a foot wrong and draw justified criticism from the specialists, but I take as my motto the words of the philosopher Donald Davidson, who once remarked, in a very different context, "Where there are no fixed boundaries only the timid never risk trespass."[3] I do not pretend to have said the last word about war and human nature. In fact, there are many important aspects that I will mention only in passing, or even ignore entirely. This book tells *part* of the larger story of war. An ocean of ink is spilled every year on agonizingly trivial questions, and trivial questions beget trivial answers. In this book I have tried to set out what I think are *the questions really worth asking*. These are "big" questions that just cannot be addressed meaningfully without engaging a certain amount of speculation. Now, speculation has its hazards. The farther you travel beyond the seemingly hard data in front of you, the easier it is to be led astray by your biases. However, when it comes to a subject as important as war, I would rather risk being wrong in a big way than timidly trying to be right in a small way. For whatever the outcome, I will have at least meaningfully extended the conversation. But this is for the reader to judge. Whatever the verdict, I hope that both my questions and the answers that I tentatively offer will contribute, however slightly, to a more restrained and reflective attitude toward war as we move deeper into this ever more precarious century.

We used to wonder where war lived, what it was that made it so vile. And now we realize that we know where it lives, that it is inside ourselves.

—ALBERT CAMUS, *NOTEBOOKS*, VOL. 3, ENTRY FOR SEPTEMBER 7, 1939

Number on deck, sir, forty-five . . . highly motivated, truly dedicated, rompin', stompin', bloodthirsty, kill-crazy United States Marine Corps recruits, SIR!

—U.S. MARINE CHANT, PARRIS ISLAND

I gave up my soul. I came back as an amputee, but you can't see my amputation. My amputation is up here, in my head, and no one can give it back to me. I just live every day and wonder if I am going to die, if the suffering will stop.

—SECOND GULF WAR VETERAN HEROLD NOEL, PFC, INTERVIEWED BY YVONNE LATTY

THE MOST DANGEROUS ANIMAL

1

A BAD-TASTE BUSINESS

> Evil is unspectacular and always human, and shares our bed and eats at our own table.
>
> —W. H. AUDEN, *HERMAN MELVILLE*

ON MONDAY, SEPTEMBER 20, 2004, Islamic militants in Iraq executed an American construction worker named Eugene Armstrong. Four men, masked and clothed in black, tensely clutched their automatic weapons while the bound and blindfolded Armstrong knelt in front of them. "God's soldiers from Tawhid and Jihad were able to abduct three infidels of God's enemies in Baghdad," the leader intoned, ". . . by the name of God, these three hostages will get nothing from us except their throats slit and necks chopped, so they will serve as an example." The long knife sliced through Armstrong's flesh. He screamed. Blood gushed from his neck. His body shuddered and became limp. The executioner placed the dripping, severed head on the back of Armstrong's lifeless body. Do you think that this is a shocking image? When the video was broadcast on national television, it was interrupted before any blood was spilled. Perhaps this discretion was a good thing; the image was, after all, very disturbing. But perhaps it would have been better to show it. Armstrong's execution was an act of war, and war is terrible. Like many terrible things, it is something that we do not want to think about too much if we can help it.

Many people conceive of war in terms of manly, granite-chinned heroes duking it out with the forces of evil. The reality is very different from this comic-book picture. It is something from which we collectively avert our gaze. The news and entertainment media obligingly maintain our illusions, protecting our sensibilities from too potent a dose of reality. This is why, during the dark days of the Cambodian genocide, the Associated Press rejected photographs of a smiling soldier eating the liver of a Khmer Rouge fighter whom he had just gutted and a soldier lowering a human head by the hair into a pot of boiling water. And it is why U.S. newspapers avoided British photographer Kenneth Jarecke's photograph of the charred head of an Iraqi soldier who was among those burned alive on Mutla Ridge during the closing chapter of the First Gulf War.[1] When the British journalist Martin Bell reported on the war in Bosnia, he quickly realized that he was expected to sacrifice reality to "good taste." The version of the war presented to television audiences was, he remarked, "about as close to reality as a Hollywood action movie," later remarking that "in our desire not to offend and upset people, we were not only sanitizing war but even *prettifying* it. . . . But war is real and war is terrible. War is a bad taste business."[2]

The cosmetic transformation of war is nothing new. Painters of the eighteenth and nineteenth centuries denuded war of its horror, portraying soldiers with "neatly-bandaged head-wounds" and "manly and heroic expressions." Much the same is true of popular literature. The literary misrepresentation of war is exemplified by the writings of Rudyard Kipling, who, despite having no combat experience himself, confidently portrayed war in ludicrously glowing terms. Kipling's romantic fantasies lured a generation of young men to their deaths in the trenches of World War I. With the advent of photography in the early nineteenth century, representations of war took on a new dimension of realism. But even the early photographers of the American Civil War, Spanish-American War, and World War I were not above manipulating things to suit their expectations. They dragged bodies into position before snapping them and passed off cleaned-up reenactments of military engagements as the genuine article.[3]

The advent of moving pictures opened new vistas for dishonestly representing battle. Most filmmakers steer safely clear of the horror and degradation of war, and most film viewers have no direct experience to act as a corrective to the Hollywood version. Consequently, many of us have an extremely distorted picture of combat. A real battlefield is not much like the typical movie version. "On the screen," writes General Sir John Hackett, "there are particular conventions to be observed."

> Men blown up by high explosives in real war . . . are often torn apart quite hideously; in films there is a big bang and bodies, intact, fly through the air with the greatest of ease. If they are shot . . . they fall down like children in a game, to lie motionless. The most harrowing thing in real battle is that they usually *don't* lie still; only the lucky ones are killed outright.[4]

Life imitates art, and the glorification of war in modern cinema has had serious consequences for the lives of its consumers. Amazingly, many young men chose to join the U.S. Marines during the Vietnam War under the influence of John Wayne films. In their minds, going to war was like being a character in a movie: good guys killing bad guys, cowboys killing Indians. In fact, during the first four months of 1968, sixty U.S. soldiers in Vietnam died trying to outdraw one another just as they had seen actors do in cowboy films.[5]

Because of these and other compelling illusions about war, it is easy—in fact, all too easy—to regard the perpetrators of mass violence as depraved monsters or madmen. For example, George W. Bush proclaimed that he ordered the invasion of Iraq and toppled Saddam Hussein's regime because he "was not about to leave the security of the American people in the hands of a madman." French president Jacques Chirac described Osama bin Laden as "a raving madman," while British foreign secretary Jack Straw described Bin Laden as "psychotic and paranoid."[6]

What evidence was there that these people were insane? There is usually none at all. The psychologists who painstakingly sifted through data on the senior Nazi officers brought to justice in the Nuremberg trials found that "high-ranking Nazi war criminals . . . participated in atrocities without having diagnosable impairments that would account for their actions."[7] They were "as diverse a group as one might find in our government today, or in the leadership of the PTA."[8] If the Nazi leaders were not deranged, what about the rank and file who did Hitler's dirty work? What about the members of the *Einsatzgruppen*, the mobile killing units that committed atrocities like the mass killing at Babi Yar, where 33,000 Jews, as well as many gypsies and mental patients, were machine-gunned to death during two crisp autumn days in 1941? Do you think that these men must have been psychopaths or Nazi zealots? If so, you are wrong. There is not a shred of evidence to suggest that they were anything other than ordinary German citizens. "The system and rhythm of mass extermination," observes journalist Heinz Hohne, "were directed by . . . worthy family men." The men of the German Reserve Police Battalion 101, a killing squad in Poland who were involved in the shooting of at least 38,000 Jews and the deportation of a further 83,000 to the Treblinka death camp, were ordinary middle-aged family men without either military training or ideological indoctrination. "The truth seems to be," writes social psychologist James Waller, "that the most outstanding characteristic of perpetrators of extraordinary evil lies in their normality, not their abnormality."[9] Purveyors of violence, terrorists, and merchants of genocidal destruction are, more often than not, people who fit the profile that Primo Levi painted of his Nazi jailers at Auschwitz: "average human beings, averagely intelligent, averagely wicked . . . they had our faces." To Hannah Arendt they were "terribly and terrifyingly normal."[10] They could be your neighbors, parents, or children. They could be you.

This book is about where war lives in human nature. It is not only, or even primarily, about people like Hitler, Stalin, or Saddam: It is about people like you and me, our ancestors, our children, and our children's children. It tells the story of why human beings, *all* human beings, have the potential to be hideously cruel and destructive to one another. Other animals attack and sometimes kill members of their own kind, but they do not organize themselves into groups to destroy neighboring communities. Mark Twain made this point over a century ago.

> Man is the only animal that deals in that atrocity of atrocities, war. He is the only one that gathers his brethren about him and goes forth in cold blood and calm pulse to exterminate his kind. He is the only animal that for sordid wages will march out . . . and help to slaughter strangers of his own species who have done him no harm and with whom he has no quarrel. . . . And in the intervals between campaigns he washes the blood off his hands and works for "the universal brotherhood of man"—with his mouth.[11]

In fact, Twain's portrait of human nature is far too charitable. Men not only march out to slaughter their own kind on a scale so huge that it beggars the imagination, they often do so in ways that are diabolically cruel. A small taste of this side of human nature is conveyed in the following passage from Thomas Alfred Walker's classic *History of the Law of Nations.*

> When Basil II (1014) could blind fifteen thousand Bulgarians, leaving an eye to the leader of every hundred, it ceases to be a matter of surprise that Saracen marauders should thirty years later be impaled by Byzantine officials, that the Greeks of Adramyttium in the time of Malek Shah (1106–16) should drown Turkish children in boiling water, and that the Emperor Necephorus (961) should cast from catapults into a Cretan city the heads of Saracens slain in the attempt to raise the siege, or that a crusading Prince of Antioch (1097) should cook human bodies on spits to earn for his men the terrifying reputation of cannibalism.[12]

Walker was writing about events that unfolded long ago and far away and that are safely confined to the pages of history books. But others that occurred before and after them equal these horrors. The screams of Basil's victims mingle with the screams of the men, women, and children whose mute remains lie in prehistoric burial sites. They mingle with the cries of the residents of Babylon when King Sennacherib's Assyrian warriors put them to the sword, and the cries of Native Americans cut down by Spanish steel and American lead. Victims of the Armenian genocide, the Jewish Holocaust, and a million other brutalities join them in a tortured chorus that echoes through history, but to which we turn a deaf ear. These facts are an embarrassment. They deflate our pretensions to moral superiority, our conception of ourselves as standing at the pinnacle of creation. Consequently, we prefer fairy tales, turning reality on its head to keep the truth at a reassuringly safe distance.

Like it or not, war is distinctively human. Apart from the raiding behavior of chimpanzees, which I will describe in chapter 4, and the so-called wars prosecuted by certain species of ant, there is nothing in nature that comes anywhere near approximating it. Despite this, we often describe warfare as "brutal" (literally "animal-like") or "inhuman"—conceiving of it as something remote from our true humanity. Another distancing tactic is to treat war as a social illness, a deviation from the naturally peaceable state of humankind, a strange cyclic malady like a fever that causes us to periodically shed the garments of civilization and fall prey to the wild beast within (Leonardo da Vinci called it "*bestialissima pazzia*," "the most bestial madness"). Even the extraordinary war correspondent Martha Gellhorn, a brave and insightful woman who reported on most of the major military conflagrations of the twentieth century, fell into this trap when she characterized war as "a malignant disease, an idiocy, a prison."[13] However much one might like it to be true, the first member of Gellhorn's trilogy of metaphors is false. War is not a pathological condition; it is normal and expectable. It is, as she went on to remark, "our condition and our history" and has been so for tens of thousands, and perhaps hundreds of thousands or even millions, of years. War is not antithetical to civilization, the brotherhood of man, or the great spiritual and cultural traditions of East and West. It is deeply and perhaps inextricably bound up with them.

What about Gellhorn's second metaphor? Is war an idiocy? Yes, in the sense of being a hideously costly way to settle conflict. Forget about the sanitized images of combat churned out for public consumption. War is grotesque. "You tripped over strings of viscera fifteen feet long," wrote William Manchester in *Goodbye, Darkness,* his memoir of World War II, "over bodies which had been cut in half at the waist. Legs and arms, and heads bearing only necks, lay fifty feet from the closest torso."[14] Guy Sajer, who fought for the Wehrmacht on the Russian front during World War II, makes no concessions to delicacy. Unlike many, he did not feel the need to put a noble gloss on unspeakable horror when he describes in his memoir *The Forgotten Soldier* how tanks grind human flesh into the dirt to make a bloody paste, their massive treads plastered with pieces of human bodies. His accounts read like descriptions of scenes from hell.

> There is nothing but the rhythm of explosions, more or less distant, more or less violent, and the cries of madmen, to be classified later, according to the outcome of the battle, as the cries of heroes or of murderers. And there are the cries of the wounded, of the agonizingly dying, shrieking as they stare at a part of their body reduced to pulp, the cries of men touched by the shock of battle before everybody else, who run in any and every direction, howling like banshees. There are the tragic, unbelievable visions, which carry from one moment of nausea to another: guts splattered across the rubble and sprayed from one dying man onto another; tightly riveted machines ripped like the belly of a cow which has just been sliced open, flaming and groaning.[15]

War is mangled bodies and shattered minds. It is the stomach-churning reek of decaying corpses, of burning flesh and feces. It is rape, disease, and displacement. It is terrible beyond comprehension, but it is not *senseless*. Wars are purposeful. They are fought for resources, lebensraum, oil, gold, food, and water or peculiarly abstract and imaginary goods like God, honor, race, democracy, and destiny.

Later on, I will argue that self-deception is an indispensable element of war, and that despite the fact that wars are calculated and planned, there is a sense in which human beings *do not know what they are doing* when they cut one another down on the battlefield. A smokescreen of self-deception is required to make most human beings capable of such acts of slaughter.

Gellhorn's third and final metaphor is both apt and powerful. War *is* like a prison. We seem to be in bondage to war. No matter how vehemently we condemn it, how forcefully we repudiate it, we are unable to free ourselves from its morbid attraction. Prisons constrain us from the outside, but there is something about human nature, something *inside* of us, that binds us to war. And so, year after year, century after century, millennium after millennium, the tragedy drags on.

Like all living things, *Homo sapiens* possess an ancient heritage; over the course of many millions of years, the forces of evolution have honed and sculpted our minds and bodies, and this patrimony has an enormous impact on how we live our lives today. The genetic programming bequeathed to us by our ancestors has many constructive, life-affirming aspects. It incites us to seek attractive mates, to savor the flavor of nourishing food, to nurture our children, to understand and control the world around us, and even to compose exquisite music and create wonderful works of art.[16] But our evolutionary legacy also has a much more disturbing face: it moves us to kill our fellow human beings. Violence has followed our species every step of the way in its long journey through time. From the scalped bodies of ancient warriors to the suicide bombers in today's newspaper headlines, history is drenched in human blood.

WHAT IS WAR?

"War" is an ordinary, workaday word and, as such, it partakes of all of the ambiguity and vagueness of ordinary speech. Such words are practical tools for dealing with the often puzzling world around us, and what they lack in precision they make up for in flexibility. Like a screwdriver, which can also be used to pry open a can of paint or lift a door from its hinges, words like "war" have varied uses. We should therefore not expect to find cut-and-dried lines of demarcation between war and other forms of violence.

There is a danger in making one's concept of war too broad, because this attenuates its meaning. Some philosophers have conceived of war as a universal principle, a dynamism pulsing at the heart of nature. The ancient Greek Heraclitus thought of war as the violent clash between opposites and the driving force behind all change, both natural and social. Many centuries later, the Arabic word "jihad" was used to denote two very different kinds of holy war: armed struggle against the infidel and the inner struggle for spiritual perfection. In the seventeenth century, Thomas Hobbes proposed that human beings living in a world of finite resources are doomed to come into conflict with one another in their effort to get a slice of the pie. Hobbes thought that antagonism simmers beneath the surface of all human interactions, constantly threatening to erupt into lethal violence, and that the problem lay in human equality. Most thinkers have regarded equality as a good thing, but Hobbes gave egalitarianism a novel twist. "From this equality of ability," he wrote, "ariseth equality of hope in the attaining of our Ends.

> And therefore if any two men desire the same thing, which neverthelesse they cannot both enjoy, they become enemies; and in the way to their End, . . . endeavour to destroy, or subdue one another. . . . If one plant, sow, build, or possesse a convenient Seat, others may probably be expected to come prepared with forces united, to dispossesse, and deprive him, not only of the fruit of his labour, but also of his life, or liberty.[17]

It follows that the basic human situation is "a condition of Warre of every one against every one."[18]

Hobbes equated war with interpersonal conflict in general. In everyday speech, we rarely use "war" nonmetaphorically in such a broad sense, but we often conflate it with murder. Our soldiers are said to have been "murdered" by the enemy (although, significantly, we are loath to claim that our soldiers murdered the enemy). However, it is misleading to treat war as murder writ large. There are important differences between them. The most significant of these, upon which a great deal turns, is that murder is an *individualistic* act whereas war pits whole *groups* against one another. This is the outward expression of something less tangible, but no less significant. Murder and war typically flow from different motives. Murder is overtly antisocial, an individualistic defiance of the state's monopoly on violence, and is typically motivated by emotions like hate, envy, greed, fear, jealousy, and spite.[19] In contrast, war is inspired—or at least justified—by "higher" purposes. It is conceived of as altruistic rather than self-centered. (We say that soldiers "serve" their country or "give their lives" for their country.) Murderers are usually moved to kill for personal reasons, while soldiers kill ostensibly in the service of an ideal of an ethical, religious, or political nature (God, freedom, democracy, etc.). The warrior typically kills to rid the world of evil, even though his* acts may entail crimes against humanity.[20] Consider the remark of Ayatollah Khomeini of Iran made during the Iran-Iraq war, in a speech in honor of Muhammad's birthday: "If one permits an infidel to continue in his role as a corrupter of the earth, his moral suffering will be all the worse. If one kills the infidel . . . his death will be a blessing to him. For if he remains alive, he will become more and more corrupt." The infidels to whom Khomeini referred were secular Iraqis. He continued:

> War is a blessing to the world and for all nations. It is God who incites men to fight and to kill. The Koran says, "Fight till all corruption and all rebellion have ceased." The wars that the Prophet led against the infidels were a blessing for all humanity. . . . A religion without war is

*I use the masculine pronoun here and elsewhere because the overwhelming majority of soldiers are male. See chapter 4 for a discussion of the gendering of war.

> an incomplete religion. . . . A prophet is all-powerful. Through war he purifies the earth. The Mullahs with corrupt hearts who say that this is contrary to the teachings of the Koran are unworthy of Islam. Thanks to God, our young people are now, to the limits of their means, putting God's commandments into action. They know that to kill the unbelievers is one of man's greatest missions.[21]

Khomeini fanned the flames of religious devotion, and his willing executioners put piety into action by forcing thousands of children to participate in human-wave assaults against Iraqi infantry, who mowed down their frail bodies. They roped children together and forced them to stumble en masse across Iraqi minefields to clear the way for advancing Iranian infantry. Each boy was issued a plastic key that he was told would open the gates of paradise.[22]

Just as too broad a concept of war entails excessive vagueness, too narrow a concept risks spurious precision. To understand war we must have an appreciation of its variety: the sometimes dramatically different forms that it has taken from time to time and place to place. To most of us, the word conjures up scenes of massed armies locked in combat, Greek infantry encircling the Persian forces at Marathon, Union and Confederate troops facing off at the Battle of Gettysburg, the slaughter on the Normandy beaches on D-Day. These were all *battles*: large-scale military engagements undertaken by mutual agreement between contending parties (1960s antiwar posters asked, "What if they gave a war and nobody came?").[23] Battle has been the main form of military activity in complex societies for many centuries, and many people equate it with war. But if we shift our focus and view war through the panoramic lens of social evolution, we encounter a very different prototype. Until the invention of agriculture, only ten thousand years ago, human beings lived in small, nomadic communities that survived by foraging for edible plants and hunting for game. Remnants of this ancient, preagricultural way of life still exist in the fast-disappearing pockets of aboriginal cultures scattered over the most inhospitable regions of the globe. These offer tantalizing clues about how our own prehistoric human beings may have lived. Of course, they are not fossilized museum pieces, relics of the distant past preserved intact in the present. Primitive cultures have undergone many transformations over time, and each possesses a unique, individual character, but there are broad similarities between many of these groups that may throw light on prehistoric ancestors' ways of life, including their ways of war.

Formal battles are a feature of tribal warfare, but these are often nothing like the kind of battles familiar to us. They are ritualistic affairs—mock battles if you like—in which men fire arrows and hurl spears at each other from a safe distance. Serious injuries and fatalities are consequently rare. A good example, described by Clarence Ray Carpenter, is a battle between two groups of Dani, a people living in the highlands of New Guinea. Carpenter tells us that neighboring groups of Dani were nearly always at war with one another and that "small groups of warrior-leaders decided when and where to fight the next round of perpetual cycles of battles." Early in the morning the men of each side prepared for the day's engagement, eating large quantities of sweet potatoes for strength and endurance. They greased themselves with pig fat, ornamented themselves with the plumes of birds-of-paradise, feathers from egrets, cassowaries, and parrots, and decorated their skin with colored clay. By noon the warriors assembled at the preordained location and then, after a series of ceremonial thrusts, "stylized advances from both sides moved, stopped, ran, stopped, advanced, and retreated, but not entirely back to where the last action started. . . . Arrows flew, signaling readiness to engage in formal conflict. There were several more advances and retreats with symbolic releases of arrows." After half an hour of these preliminaries, the battle was joined in earnest.

> The open space is filled with cavorting men in motion who threw and dodged flights of spears and arrows. Massed men in groups of 30 to 50 charged near together in the mid-ground to the rhythmic sound of several hundred feet on hard ground and clashed in action. Spears and arrows were

> discharged, flew in waves and were avoided for an exhausting hot half hour.

As the afternoon wore on, the action began to wind down. The men living farthest away departed first, so they could make it home before darkness fell, and as the ranks thinned, the fighting degenerated into verbal taunts and threats. "These tirades," writes Carpenter, "hurled across the 100-yard-wide zone of conflict gave the weary warriors pleasure, released their tension, and were amusing. Personal insults were shouted and laughter resulted."[24]

So far, primitive warfare sounds like a relatively innocuous affair. However, ceremonial battle is not all there is to primitive warfare; there is also a more lethal practice called "raiding." In a raid, a small band of men stealthily enters enemy territory with the purpose of killing any men and abducting any young women whom they encounter. Raids minimize risk to the attackers. They target people who are relatively defenseless. Fatal wounds are typically inflicted as the ambushed victims try to flee.[25] Anthropologists believe that raiding was probably the form of warfare prosecuted by our prehistoric ancestors. One of the best descriptions of it is a detailed, firsthand narrative by a woman named Helena Valero, who was abducted by Yanomami Indians and lived among them in the Venezuelan rain forest for twenty-four years. Her memoir begins with a description of how Helena and her parents were attacked by a band of Yanomami. Her mother and father were wounded by arrows and fled, but the thirteen-year-old Helena was captured and absorbed into the Yanomami community. Some time after these events (we are not told how long) an old man arrived at the village to warn of an impending attack by a group called the Karawetari. Men immediately began to build a stockade of tree trunks to protect the *shapuno* (village) from assault, and the chieftain posted five sentinels. Several days later, Helena wandered into the forest with a young woman and a child to gather fruit. Suddenly, she heard shouts of "The enemy! The enemy!" and noticed women and young girls running frantically.

> The mother took the child from my knees, gave me a push and said: “Let’s run! Let’s run!” So we all ran towards the *shapuno*; we found it almost empty. The men had gone running to meet the enemy; the other women, with their children, had run away.

Helena caught up with the other women and children and sat down to rest, but soon a messenger arrived to warn them that the enemy was rapidly approaching. They fled again, running all day. When night fell, two men arrived and advised them to keep moving, so they picked up torches and ran all though the night. They eventually took refuge on a hillside, but the Karawetari raiders had spotted them. Dividing in two, one contingent climbed the hillside to a point above where the women and children were hiding, while the others remained below.

> Those who had climbed up the hill and were above us began to shoot arrows. . . . Six or seven arrows fell, but did not hit us. A child, trembling with fear, climbed up a tree. . . . The enemy were coming down from above and climbing up from below. The boy who had climbed a tree shouted at a man who was coming nearer: “Father, don’t shoot me!” “I’m not your father,” shouted the man . . . and he shot him. The arrow hit the little boy from behind in the leg and came through in front. The child fell, picked himself up and ran with the arrow still in him.

Next, the men of the raiding party began to seize the women and kill their male children.

> One woman had a baby girl in her arms. The men seized the little child and asked: “Is it a boy or a girl?” and they wanted to kill it. The mother wept: “It’s a little girl, you mustn’t kill her.” Then one of them said: “Leave her; it’s a girl; we won’t kill the females. Let’s take the women away

> with us and make them give us sons. Let's kill the males instead." Another woman had a baby boy only a few months old in her arms. They snatched him away from her. . . . They took the baby by his feet and bashed him against the rock. His head split open and the little white brains spurted out on the stone. They picked up the tiny body, which had turned purple, and threw it away. I wept with fear. . . . Meanwhile from all sides the women continued to arrive with their children, whom the other Karawetari had captured. They all joined us. Then the men began to kill the children; little ones, bigger ones, they killed many of them. They tried to run away, but they caught them, and threw them on the ground, and stuck them with bows, which went through their bodies and rooted them to the ground. Taking the smallest by the feet, they beat them against the trees and the rocks. The children's eyes trembled. Then the men took the dead bodies and threw them among the rocks, saying: "Stay there, so that your fathers can find you and eat you." . . . All the women wept.[26]

Many scholars hold that raiding is not "true" warfare. "Primitive warfare" is said to exist "below the military horizon" and it must be sharply demarcated from real, "civilized" war.[27] There is clearly something wrong with this picture. Drawing a hard boundary between primitive and civilized warfare is an arbitrary move that throws obstacles in the path of anyone trying to understand the relationship between war and human nature. Nature abhors discontinuity. These concepts may be useful for slicing the pie of reality into bite-sized portions, but they are ultimately just a convenience—simplifying fictions that we impose upon the world to help us find our way through its complexities. They do not, as Plato put it, "carve nature at its joints," but hack off parts "like a clumsy butcher."[28] That being said, there is also something valuable about the distinction. Raids are unilateral surprise attacks characterized by large imbalances of power. Many people attack a few, and there is little or any risk to the attacker. Battle is bilateral, and soldiers in both camps face the possibility of injury or death.[29] The distinction between raiding and "true war" is also reflected in their chronology. "True war" is a relatively recent development; as far as we know, it began in the Middle East around ten thousand years ago. Raiding is much more ancient, so ancient, in fact, that it probably predates the hominid line.

A naturalistic approach to war should conceive of it in the same way that we conceive of the rest of nature: as a continuous whole. To understand war, we must resist the temptation to box it in tidy categories hedged about with arbitrary distinctions. War encompasses a wide spectrum, from raids in the Amazon rain forest, through gang battles on the streets of Los Angeles to the nuclear annihilation of Hiroshima and Nagasaki. Although these phenomena are very diverse, it is not too difficult to bring all of them together under a single umbrella: *War is premeditated, sanctioned violence carried out by one community (group, tribe, nation, etc.) against members of another.*[30] It is worthwhile to spend a little time unpacking all of this. The expression "violence" is intentionally vague. What counts as violence? Killing certainly does, but so, too, does rape, injury, or coercion based on the threat of death or injury. Although wars are nearly always characterized by slaughter, "collective killing for some collective purpose," as military historian John Keegan puts it, not every act of war is an act of killing.[31] The term "community" is also indeterminate, and I will have a lot more to say about it in chapter 9. For now, let's note that communities can be as small as two or three individuals or as large as thousands or even millions of people. Warring communities are rivals, and rivalry implies *conflicts of interest.* People fight *over* something. They fight over resources like territory, food, and water. In the case of religious or ideological wars, they may fight over less tangible commodities like souls (in the contemporary parlance, "hearts and minds"), honor, or justice. Sometimes—perhaps often—perceptions of rivalry do not correspond to reality. During the early 1940s the national government of Germany waged a genocidal campaign against the Jews of Europe based on the false premise that they were fighting against an international conspiracy to control the world.

In war, individuals are attacked because of their community affiliation. In other words, although wars kill individuals, they are *between* groups; an individual human being is an enemy only insofar as he or she is a representative of a group, be it tribal, religious, national, ideological, or whatever.

Some readers may feel that definition of war is far too broad. True, it goes far beyond what we normally mean by "war," but there is method in my madness. An expansive definition helps to clarify how war fits into the larger fabric of nature. We need to stake out a wide-ranging conceptual territory, emphasizing the continuity rather than the discontinuity between diverse phenomena, looking beyond their varied culturally and historically specific manifestations to grasp their common core.

TERRORISM, ATROCITY, AND GENOCIDE

"Terrorism," "atrocity," and "genocide" are words for the dirty side of war. Unfortunately, they are at best vague, and at worst meaningless. They are obstacles to getting a clear perspective on war. In this section, I want to spend a bit of time considering their uses and abuses.

Ever since 9/11, it is has become customary to sharply distinguish war from terrorism. However, there is no universally accepted definition of terrorism. According to one definition accepted by the United States government, war is military action by a nation-state, whereas terrorism is "premeditated, politically motivated violence perpetrated against noncombatant targets by subnational groups or clandestine agents."[32] This definition not only presupposes a restrictive definition of war as an activity of nation-states, it also implies that *the very same act* of aggression that counts as "war" when overtly implemented by a national government is "terrorism" if carried out covertly or by a subnational group. This is unblushingly arbitrary: a distinction of convenience, rather than of substance.

What about noncombatant targets? It certainly sounds nice to speak about civilian deaths as merely occasional by-products of soldierly shoot-outs, but this idea is far removed from the truth. Wars kill *more* civilians than they do combatants. Of the 87,500,000 or so deaths in twentieth-century wars, around 54,000,000, or just over 60 percent, were civilians. This percentage is an average; some wars kill less, but others kill far more (during the 1990s, 75 to 90 percent of the world's war fatalities were civilians).[33] Under international law, civilian deaths are permissible if they are "proportional" to the anticipated military advantage (however, what counts as an acceptable proportion of civilian deaths is left unspecified).[34] Former war correspondent Chris Hedges points out that noncombatants are routinely "shot, bombed, raped, starved and driven from their homes." Displaced people typically suffer from hunger, disease, and hideously elevated mortality rates.

> In 2001, 40 million people were displaced from their homes because of armed conflict or human rights violations. . . . Five million Europeans were uprooted from 1919 to 1939. World War II displaced 40 million non-Germans in Europe, and 13 million Germans were expelled from countries in Eastern Europe. Approximately 2.5 million of the 4.4 million people in Bosnia and Herzegovina were driven from their homes during that region's war in the early 1990s. More than 2 million Rwandans left their country in 1994. In 2001, 200,000 people were driven from Afghanistan to Pakistan.[35]

Incidentally, many of these fatalities are children.

> More than 2 million children were killed in wars during the 1990s. Three times that number were disabled or seriously injured. Twenty million children were displaced from their homes in 2001. Many were forced into prostitution. A large percentage of those will contract AIDS.[36]

Children are also combatants, and over 40 percent of the armed organization worldwide make use of them. Around two million were killed or wounded in the last decade, and many others remain psychologically scarred for life. Today, there are around 300,000 child soldiers in the world. Many of them are abducted and forcibly recruited, but others join as a way to survive in dangerous, war-torn regions. This military is nothing new. It has been practiced continuously from remote antiquity to the present, all over the world. Even in the American Civil War around 100,000 soldiers were less than sixteen years old.[37]

So, it is wise to ignore the pious pronouncements of the politicians and news media. The grim realities of war are easily concealed behind faceless statistics and euphemistic jargon. The situation on the ground is always stomach churning. Innocent people *always* die.

A different approach defines terrorism as acts of aggression performed to intimidate a population or a government. However, this will not do as a criterion separating terrorism from war because inculcating fear is a standard military procedure. It is the strategy of "shock and awe," which, according to its creator, Harlan Ullman, "rests ultimately in the ability to frighten, scare, intimidate and disarm."[38]

"Terrorism" is a word with little content—it is a label for brutalities committed by "the enemy," and from which one's own acts of destruction are exempted. It is an inchoate and emotionally laden concept, a semantic mirror of our dishonesty and a repository for everything about war that we would like to disavow. Making a sharp distinction between war and terrorism is at best a self-deceptive game.

The word "atrocity" is like "terrorism" (with which it is often coupled) in that both generate more heat than light. The more precise equivalent used in international law is "crimes against humanity," which are defined as the systematic and intentional murder, extermination, enslavement, deportation, imprisonment, torture, sexual violence, apartheid, or commission of other inhumane acts against civilians.[39] As with terrorism, we imagine that such acts are committed exclusively by the other side and that they are fundamentally alien and profoundly antagonistic to our own way of life. Nevertheless, "terrorist atrocities" are celebrated in the cultural traditions of many nations, including our own. The Holy Bible is full of examples. Psalm

137 is a particularly well-known and very beautiful biblical text, but it contains the following ugly line: "O daughter Babylon. . . . Happy shall they be who take your little ones and dash them against the rock!": a matter-of-fact reference to the routine practice of infanticide in ancient warfare.[40]

The concept of genocide is more meaningful than "terrorism" and "atrocity." The Russian philosopher and legal scholar Raphael Lemkin introduced it in 1944 to denote "a coordinated plan of different actions aiming at the destruction of essential foundations of the life of national groups, with the aim of annihilating the groups themselves." Lemkin went on to specifics:

> The objectives of such a plan would be disintegration of the political and social institutions, of culture, language, national feelings, religion, and the economic existence of national groups, and the destruction of the personal security, liberty, health, dignity, and even the lives of the individuals belonging to such groups. Genocide is directed against the national group as an entity, and the actions involved are directed against the individuals, not in their individual capacity, but as members of the national group.[41]

Four years later, the United Nations adopted the concept of genocide. Resolution 260 defines it as "acts committed with intent to destroy, in whole or in part, a national, ethnical, racial or religious group" such as killing, causing physical or psychological harm, "inflicting on the group conditions of life calculated to bring about its physical destruction in whole or in part," "measures intended to prevent births within the group," and, finally, "forcibly transferring children of the group to another group." Although we associate genocide with mass killing, the U.N. definition allows that genocide can occur without any loss of life. It is also restricted to acts against "national, ethnical, racial or religious" groups but *excludes* mass violence against political or ideological groups. These omissions and inconsistencies led University of Hawaii political scientist R. J. Rummel to introduce the term "democide" to cover *all* forms of politically motivated government-sponsored killing apart from warfare.[42] Estimates of the death toll from twentieth-century democides range from 80,000,000 to 170,000,000 lives.[43]

Over four thousand years ago the anonymous Sumerian author of the "Lament for Unug" represented the horrors of war as a rampaging monster—an image that is as apposite today as it was then ("Unug," by the way, is the original Sumerian form of the name handed down to us as "Iraq.")

> Its mouth shall be grotesque—a blaze that extends as far as the nether world. Its tongue shall be an inferno, raining embers. . . . Its great haunches shall be dripping knives, covered with gore, that make blood flow. Its muscles shall be saws that slash, its feet those of an eagle. It shall make the Tigris and Euphrates quaver, it shall make the mountains rumble. At its reverberation the hills shall be uprooted, the people shall be pitched about like sheaves.[44]

THE RETURN OF THE KILLER APE?

One explanation for the prevalence of war is the idea that a bent for violence was bred into our species over eons of evolutionary time. This view is associated with a scientist named Raymond Dart, who argued that human beings are, at bottom, killer apes. Dart's thesis has not aged well and is often ridiculed—especially by those who reject any attempt to give an evolutionary account of human violence. My argument has some affinities with Dart's hypothesis, so it is worthwhile pausing to consider some of the latter's strengths and weaknesses.

Here's the story. One day in 1924, a medical student at the University of Witwatersrand presented her anatomy professor with a fossilized baboon cranium recovered from a limestone quarry in a place called Taung in South Africa's North West Province. The professor, a neuroanatomist named Raymond Dart, had an interest in human origins, and it occurred to Dart that a site yielding prehistoric baboon remains might also be a source for bones of ancestral human beings. He asked a geologist colleague who was planning to visit Taung to see what he could find. Before long a crate of fossils arrived on Dart's doorstep, and it included an unusual-looking skull partially encased in limestone. Dart spent two and a half months painstakingly chipping away the encrustations to reveal a magnificent find. The skull had once belonged to a three-year-old child of an extremely ancient human-like creature that Dart named *Australopithecus africanus* (literally, "southern African ape").

The discovery sharply contradicted prevailing scientific opinion about where the hominid line began, and was initially treated with scorn. True, Darwin had hypothesized that the human species originated in Africa, but in the 1920s the weight of scientific evidence pointed toward Europe and Asia as the most likely birthplace for humanity. In the early nineteenth century several skull fragments had been discovered in Belgium, and in 1848 a complete specimen came to light in Gibraltar. Next, in 1856, a sensational find was made in the Neander Valley in western Germany, and was appropriately named *Homo neanderthalensis*, or "Neanderthal Man." In the later part of the century, attention shifted to Asia as prehistoric skeletal remains began to turn up on the island of Java (these were later identified as coming from *Homo erectus*, one of our earliest direct ancestors). Then, between 1908 and 1917, several bone fragments were "discovered" in a gravel pit in southeast England, and although the discovery was later exposed as a hoax, it reinforced the view that human beings originated in Europe.[45] However, by the 1940s the discovery of several more African specimens turned the tide of scientific opinion in Dart's favor, and the idea that the human lineage began in Africa became widely if not quite universally accepted.

Dart originally (and probably correctly) thought that *Australopithecus africanus* was a gentle fruit eater that supplemented its diet by hunting small game and birds' eggs. However, he later changed his mind and came to believe that australopithecines were big-game hunters, speculating wildly that these creatures also turned their predatory appetites against one another, brandishing animal bones as deadly weapons and feasting on one another's raw flesh. "The loathsome cruelty of mankind to man," he wrote in famously overwrought prose, ". . . is explicable only in terms of his carnivorous, and cannibalistic origin."

> The blood-bespattered, slaughter-gutted archives of human history from the earliest Egyptian and Sumerian records to the most recent atrocities of the Second World War accord with early universal cannibalism, with animal and human sacrificial practices or their substitutes in formalized religions and with the world-wide scalping, head-hunting, body-mutilating and necrophilic practices of mankind in proclaiming this common bloodlust differentiator, this predaceous habit, this mark of Cain that separates man dietetically from his anthropoidal relatives and allies him rather with the deadliest of Carnivora.[46]

In the late 1970s, research by C. K. "Bob" Brain questioned Dart's interpretations of the skeletal evidence. Brain demonstrated that the animal bones that Dart thought were primitive weapons bore all the earmarks of the remains left by leopards after a predatory feast, and the jagged holes that he found in ancient skulls, and which he assumed were inflicted by weapon-wielding fellow australopithecines, were probably left by leopard teeth (recent research suggests that the Taung child was carried off and killed by a huge bird of prey whose talons pierced its fragile skull). The heaps of bones that Dart took to be evidence for primal cannibalism were reinterpreted as more likely remains left by predators whose menu included both australopithecines and four-legged prey. *Australopithecus* was neither a bold hunter nor a vicious cannibal, but a weak and furtive scavenger, nervously looking over its shoulder in fear of hyenas, lions, and saber-toothed tigers, for whom it was easy prey.[47]

Some students of the human condition take Dart's error to imply that human beings are not innately aggressive. For example, Donna

Hart and Robert W. Sussman kick off a recent book on human prehistory by heaping ridicule upon the killer-ape theory.

> Does the *Homo sapiens* flower at the end of the killer-ape stem still contain dark seeds from ancestors who slew giant beasts with rocks and bashed each other's heads in during routine bloody torrents of violence? Well, that's a very *convenient* excuse for the unpleasant aspects of our human condition, but evidence points to quite contrary circumstances.[48]

There is little doubt that our ancestors were hunted by larger, stronger, more ferocious predators, and that their long experience of being hunted, generation after generation, for millions of years, played an important role in configuring the human mind. They survived by foraging for plants and small animals and scavenging what meat they could from the partially eaten carcasses left by large carnivores. But it is equally true that they eventually managed to turn the tables on the monsters that stalked them and became the most masterful predators that the world has ever known (today, the lions and tigers need more protection from *us* than we need from them). This was an extraordinary transition, which is to the best of my knowledge, unparalleled elsewhere in nature. As we will see in chapter 10, it is vital for understanding the psychology of war. "A million years ago lions (or something very like them) were lions and antelopes were antelopes, and one the prey of the other pretty much as they are now," writes Barbara Ehrenreich. "Only the line of *Homo* made a decisive advance up the food chain, learning to band together in highly organized ways, to augment their strength with fire and sharpened stones, and to enrich their strategies with verbal memories of past exploits."[49] At some point during or immediately after this remarkable transition, war became an ongoing feature of human life and a distinctive characteristic of our species.

Hart and Susman propose that because our earliest ancestors were more hunted than hunters, this means that we are not *naturally* killers. They urge us to "quit accepting our spurious heritage as Man the Hunter to excuse why we start wars, torture others and scorch the earth."[50] In saying this, they speak for many who believe that giving an evolutionary account of war is an endorsement of the geopolitical status quo misleadingly packaged as an objective truth. They feel that it spuriously *justifies* human brutality by offering what amounts to a universal insanity defense stamped with the imprimatur of science.

Although this reasoning is specious, the landscape of assumptions spread out behind it is well worth exploring, because it points to fundamental obstacles that must be overcome if we are ever going to understand, and be in a position to do something about, the curse of war. It is instructive to compare the study of the AIDS virus with the study of human mass violence. No reasonable person would claim that investigating the evolution of HIV *endorses* the disease. So, why should an interest in the evolutionary origins of war be treated any differently? Why is it that we applaud the study of deadly infectious diseases but are indifferent, dismissive, or even hostile toward efforts to understand the lethal side of human nature? This sharp dichotomy flows from an insidious and remarkably persistent illusion about what it is to be a human being. In spite of all that we have learned about the mind over the last century or so, there is still a stubborn tendency to treat it as something that transcends nature. Thanks to careful scientific research, we now know a good deal about the neurochemistry that makes brains run, the relationship between psychological processes and specific areas in the brain, and the mechanisms underpinning learning, memory, and perception. As a result, human beings are now at a crossroads in the way that they understand themselves. Faced with an impressive and rapidly developing scientific image of the human animal, it is difficult to avoid the conclusion that our mental states—the thoughts that we think, the passions that move us, and the decisions that mould our lives—are consequences of physiological processes occurring in our brains. But it is hard for many people to abandon the concept that human beings are angels imprisoned in earthly shells. According to this ancient, emotionally compelling vision, the core of a person is their nonphysical soul—a thing that is distinct from all that is flesh and blood, and immune from the causal laws that govern the behavior of merely physical things. It follows from this that human behavior is radically different from the behavior of nonhuman life forms because we, and we alone, possess free will. Lesser beings blindly, rigidly, and mechanically act out their biological programming, whereas we humans *choose* how to live. This supposedly makes us the authors of our destiny, and this opens up a chasm between the human world and the realm of nature. Free will makes us morally responsible for our actions, but explaining human behavior biologically robs us of responsibility and reduces us to the stature of mere animals. This dread of "the Specter of Creeping Exculpation," as Tufts University philosopher Daniel Dennett wryly calls it, is badly misplaced.[51] Although some pundits assure us that human violence is best explained "by choice and not by our status as bipedal primates," the idea that we are violent by choice rather than by nature implies that choice can be divorced from nature.[52] It can't. Rabbits "choose" to eat plants, and wolves "choose" to eat meat. Their choices are based on the kinds of animals that they are, and the same principle applies to us. Human beings wage war because it is in our nature to do so, and saying that war is just a matter of choice without taking into account how our choices grow in the rich soil of human nature is a recipe for confusion. The question that needs answering is, "What in human nature causes us to choose war?" Dart gave the wrong answer to the right question. He knew that there had to be *some* evolutionary story explaining the warlike behavior of humankind, but he failed to put his finger on how and why human beings became such proficient killers. Those who in principle reject an evolutionary account of collective human violence must either deny its existence—which is surely a quixotic move—or else provide an alternative hypothesis. So far, no coherent alternative has been suggested.

In the following chapters, I will argue that war's allure comes from tendencies inscribed in our genes over evolutionary time, and that violent conflict benefited our ancestors, who were victors in the bloody struggle for survival. This is why the disposition to war lives on in us, and why we periodically yield to it and are drawn down into a hell of our own making. However, saying that war is driven by biology is not the same as saying that it is inevitable. Indeed, the most important reason for teasing out the biological roots of war is to find better ways to prevent it. We may be able to unlock the door of our prison. In many areas of life, it is possible to control our behavior by cleverly manipulating our own genetic imperatives. For example, human beings are endowed with sexual urges that are far more relentless than those of most other animals, and yet many of us are able to maintain monogamous marriages, and some even embrace celibacy. Likewise, most people are irrepressibly fond of the taste of sugar and animal fat, but we can restrict our consumption of calorie-rich foods in the interest of health or a more attractive appearance—even to the point of self-starvation. How do we manage these spectacular feats of self-engineering? Not by transcending our natural impulses. It is impossible for us to pull ourselves up by the proverbial bootstraps and elevate ourselves above our own nature. No, we control our behavior by cleverly playing nature off against itself. The desire for attractive mates and to live a long and vigorous life are every bit as "biological" as the craving for sugar and meat, and it is these desires that fuel the self-deprivation of dieting. Staying mindful of the value of a long-term mate helps us resist the temptations of casual sex. These accomplishments are often hard won, and our efforts to secure them are not always successful, but it is obvious that the more we know about ourselves, the more skillfully and effectively we can pull the strings that control our own behavior. Human beings may not be doomed to war. We may be able to break its spell and take control of our future. But to do this we must be willing to look at ourselves, and face some stark, unflattering truths.

WAR, RELIGION, AND SCIENCE

This book is premised on the idea that it is both possible and desirable to understand our capacity for war scientifically. Unfortunately, there are increasingly large numbers of people who reject this idea on religious grounds. Fundamentalism is on the march, and it has permeated our lives in serious and, I think, threatening ways.

William Jennings Bryan was a paradigmatic example of politicized Christian fundamentalism. He was an impressive figure: a three-time Democratic Party presidential candidate, secretary of state under Woodrow Wilson, and prosecuting attorney in the Scopes Monkey Trial. Bryan's religious beliefs made him critical of the growing secularism of American society. He believed that the hegemony of science spawned moral decadence, and proposed legislation making it illegal to teach "atheism, agnosticism, Darwinism, or any other hypothesis that links man in blood relationship to any other form of life" in any publicly funded educational institution."[53] Tennessee representative John Washington Butler took up Bryan's suggestion. Butler's bill banning "any theory that denies the story of the Divine Creation of man as taught in the Bible, and to teach instead that man has descended from a lower order of animals," became law in 1925. It was this law that John Scopes and the American Civil Liberties Union set out to challenge in the great trial of 1925. The trial was more than a dispute about the law: it became a showcase for a clash between fundamentalist and scientific worldviews.

Bryan held that the moral decadence promoted by evolutionary biology is evident in the character of modern warfare. "Science," he wrote in his summation, "has made war so hellish that civilization was about to commit suicide. . . ."

> And now we are told that newly discovered instruments of destruction will make the cruelties of the late war seem trivial in comparison with the cruelties of wars that may come in the future. If civilization is to be saved from the wreckage threatened by intelligence not consecrated by love, it must be saved by the moral code of the meek and lowly Nazarene. . . . A bloody, brutal doctrine—evolution—demands, as the rabble did nineteen hundred years ago, that He be crucified.[54]

Bryan was not entirely wrong. Science is a tool, and if we were an essentially kind and peaceful species, it would not occur to us to use this tool for destructive purposes. We exploit science to make war *because we are warlike creatures.* If science did not exist, we would still be killing one another. We would be crushing our enemies' skulls with stones instead of blasting them to pieces with bombs. Bryan was right to say that the immense power unleashed by scientific knowledge is dangerous unless used in a morally responsible way. But he was wrong to assert that religious belief and a rejection of the "bloody, brutal doctrine" of evolution are means to that end. Not only is there no evidence that religion makes people morally responsible, but history confronts us with a long, bloody record of wars, genocides, and other atrocities inspired by religious devotion and often executed by religious institutions. Consider the following eyewitness description of an historical event:

> Mothers were skewered on swords as their children watched. Young women were stripped and raped in broad daylight, then . . . set on fire. A pregnant woman's belly was slit open, her fetus raised skyward on the tip of sword and then tossed onto one of the fires that blazed across the city.[55]

This sounds like a nightmarish scene from the distant past—perhaps an account of barbarians sacking Rome, or an excerpt from the story of the First Crusade. In fact, it is from a report in the July 27, 2002, edition of *The New York Times* describing religiously motivated riots between Muslims and Hindus that occurred in India earlier that year. Events like this have occurred repeatedly ever since the founding of the three major monotheistic religions of Judaism, Christianity, and Islam. Believers in the One True God have racked up an impressive record of atrocities over the centuries, including the attacks on the Pentagon and World Trade Center on September 11, 2001.

I pin my hopes on science. Science may be unable to find a solution to the problem of war, but it is our best shot. This effort is beset with difficulties, both intellectual and emotional. Although we are members of a relentlessly curious species, some topics are difficult to think about because they are so disturbing. Consequently, it is all too easy to substitute pleasant fantasies for harsh realities, and to paint a flattering but fundamentally untruthful portrait of human nature and the forces that drive it. Still, we cannot afford to avert our gaze.

2

EINSTEIN'S QUESTION

> War is a horrible repetition.
>
> —MARTHA GELLHORN, *THE FACE OF WAR*

DAWN BROKE ON THE ERA of mass destruction in 1914. This was the start of World War I, touted by the mass media as the "war to end all wars." The outbreak of hostilities was greeted with feverish excitement and elation, cheers, patriotic slogans, and seas of waving banners. Four years later, when the smoke cleared, death was everywhere. More than eight million men lay butchered on the battlefields of Europe, and a further fifty million or so were wiped out by the flu pandemic that swept around the globe in its immediate aftermath. The war made a mockery of Europe's presumptions of moral progress. It was, in the words of the writer A. A. Milne (of *Winnie-the-Pooh* fame) "a degradation which would soil the beasts, a lunacy which would shame the madhouse."[1] Once the nightmare was over, the victorious allied powers established the League of Nations to promote cooperation, peace, and security in an increasingly dangerous world.

In 1932, against the ominous backdrop of gathering political instability in Germany, the League of Nations invited Albert Einstein to engage in a public exchange on any subject with a discussant of his own choosing. Einstein's reputation as a scientific genius of the first magnitude was already firmly established. Ten years earlier he had received the Nobel Prize in physics for discoveries that revolutionized our ideas about the fabric of the physical universe. Einstein was also well known as a staunch and outspoken pacifist and crusader for social justice who left no one in doubt about his passionate political views. "Heroism on command," he once said, "senseless violence, and all the loathsome nonsense that goes by the name of patriotism—how passionately I hate them! How vile and despicable seems war to me! I would rather be hacked to pieces than take part in such an abominable business."[2]

Einstein's choice of topic was consistent both with his social conscience and with the broad mission of the League of Nations: he wanted to discuss the problem of war. Instead of approaching a fellow physicist or a statesman, political scientist, or military historian, Einstein turned to the elderly Sigmund Freud, asking him to address what he described as "the most insistent of all the problems civilization has to face."

> This is the problem: Is there any way of delivering mankind from the menace of war? It is common knowledge that, with the advance of modern science, this issue has come to mean a matter of life and death for civilization as we know it; nevertheless, for all the zeal displayed, every attempt at its solution has ended in a lamentable breakdown.[3]

Einstein believed that war is so obviously irrational that we have to explore the "dark places of human will and feeling," as he put it to Freud, to make sense of it. He was perplexed by the power of propaganda to drive men to sacrifice their lives on the battlefield, and thought that this could only be explained by an innate "lust for hatred and destruction" that is inflamed by political propaganda. He beckoned Freud to "bring the light of your far-reaching knowledge of man's instinctive life to bear upon the problem."[4]

Freud was in his seventies when he received Einstein's invitation. Despite his advanced age, and the cancer that had been gradually consuming him for over a decade, Freud's intellectual fire still burned brightly. He responded hesitantly, fearing that his bleak assessment of human nature would be unpalatable to the League of Nations. "All my life I have had to tell people truths that were difficult to swallow," Freud remarked to a League official who visited him in Vienna. "Now that I am old, I certainly do not want to fool them."[5] Freud had spent the bulk of his long career searching for the forces that drive human behavior, first as a neuroscientist and later as an explorer of the dark byways of the unconscious mind. Over the course of his long career, Freud came to the conclusion that human beings are self-deceptive creatures that are moved by instinctual urges operating outside the jurisdiction of conscious awareness and control. For many years, Freud had thought that sexual desire is the most potent of these motivational powerhouses, but the unprecedented carnage of World War I caused him to reconsider his position. While continuing to embrace the idea that sex is an extremely important moving force in human behavior, Freud reluctantly came to believe that our destructive urges are equally significant, and that these twin conflicting forces drive our lives from birth to death.

The exchange between Einstein and Freud was published in 1933 under the title *Why War?* the same year that Adolf Hitler triumphantly assumed the chancellorship of Germany. The little book was banned by the Nazi regime. It was included in the book burnings by student supporters of the Third Reich and consigned to the flames with the words "On behalf of the nobility of the human soul—I offer to the flames the writings of one, Sigmund Freud." When the news reached Freud in Vienna, he commented dryly, "What progress we are making. In the Middle Ages they would have burned me. Now they are content with burning my books."[6] He could not have anticipated that within a few years three of his sisters would be incinerated in the ovens of the Treblinka death camp and that the fourth would die of starvation in Theresienstadt. He could not have known that just two decades after the "the war to end all wars" the world would embark on an unparalleled plunge into darkness that would devastate Europe and rob over fifty million people of their lives.

Freud's response to Einstein centered on the idea that we, like other animals, are inclined to resolve conflicts of interest by resorting to violence. Initially, dominance over others was a function of sheer muscular strength. Later, the intelligence required to manufacture and expertly wield weapons trumped simple brawn. Through the rule of law, whole communities could unite in suppressing violence within their jurisdiction, but this did not obliterate violent conflict *between* communities, and the achievement of in-group cohesion led to entire groups attacking one another. Freud thought that the urges fueling both individual and collective violence are based on a "death instinct," a self-destructive drive that must—in the interest of self-preservation—be directed outward as aggression toward others. Freud could not envisage any way to bring this force under control, at least in the foreseeable future. "An unpleasant picture comes to mind," he wrote, "of mills that grind so slowly that people may starve before they get their flour."[7]

Freud's response to Einstein was less than satisfactory when viewed from the privileged vantage-point of twenty-first-century science. It was not poorly reasoned or ill-informed—far from it. But Freud's efforts were hampered by the fact that the necessary scientific resources were just not available in 1932. Human evolution was barely understood, behavioral genetics and cognitive neuroscience were nonexistent, and experimental psychology was extremely crude. In just a few decades all of this would change dramatically. We now know immensely more about human behavior than during Freud's lifetime. We possess a wealth of empirical data, precise mathematical models, and a range of powerful scientific theories upon which to draw. Today, we are in a position to pick up where Freud left off and bring contemporary scientific knowledge to bear on Einstein's question.

Accounts of warfare stretch back to the dawn of history. One of the most ancient historical records in the world, the Narmer palette, which was carved around 3000 B.C., depicts the predynastic Egyptian king Narmer with one arm majestically raised, war mace in hand, poised to smash his kneeling enemy's skull. The earliest manual on military strategy dates to around 400 B.C., when the Chinese general Sun Tzu wrote his great book *The Art of War.* Writings about the morality and legality of war are almost as old. Thomas Aquinas discussed the concept of "just war" in the thirteenth century in his *Summa Theologica*, but its roots reach back to the fourth century writings of Augustine, to the Roman philosopher Cicero in the first century B.C., and perhaps even to the Song of Deborah in the Book of Judges. In contrast, the *scientific* study of war is a recent phenomenon. One might imagine that, given its incomparable importance, investigations of war would be a centerpiece of the social sciences. Not so. One telling survey of the world's three leading sociological journals revealed that less than 3 percent of the articles published between 1936 and 1984 concerned war, and almost all of these were published in 1942. A survey of articles in the same journals between 1986 and 2000 revealed that fewer than 1 percent dealt with war, and *none of them* considered the causes of war.[8] There is comparable neglect in other social science disciplines. According to Lawrence H. Keeley, professor of anthropology at the University of Illinois and the author of *War Before Civilization*, archeologists of prehistory have had difficulty coming to grips with armed conflict.

> Less by sustained argument than by studied silence and fashionable reinterpretation, prehistorians have increasingly pacified the human past. The most widely used archaeological textbooks contain no references to warfare until the subject of urban civilizations is taken up. The implication is clear: war was unknown or insignificant before the rise of civilization.[9]

Keeley's diagnosis is echoed and extended by Steven A. LeBlanc, director of collections at Harvard's Peabody Museum of Archaeology and Ethnology, who states that archaeologists and anthropologists have been blind to the prevalence and importance of war in prehistoric societies. Turning to psychology, we find an extensive research literature on violence, mainly treating it as a learned behavior, but research dealing specifically with war is thin on the ground. The psychiatric literature *does* address war, but only as a cause of mental illness (variously called post-traumatic stress disorder, shell shock, traumatic neurosis, battle fatigue, or "nostalgia").[10]

The serious study of war as a natural phenomenon is long overdue. We owe it to ourselves and to future generations. In an age dominated by the shadow of weapons of mass destruction, we have no time to waste.

THE NATURALNESS OF WAR

Einstein and Freud thought that war and its related atrocities come naturally to human beings. It hardly needs saying that this is a depressing prospect, one that we almost reflexively reject. However, the evidence that I will present to you in this book overwhelmingly supports the naturalness of war.

To make sense of this, we need to consider what it means to say that war is natural. This is not as obvious as it might at first sound. The great eighteenth-century Scottish philosopher David Hume (who will be making many more appearances in this book) pointed out in his monumental *Treatise of Human Nature* that the word "natural" has a very broad range of meanings.[11] Sometimes, he noted, we use "natural" as the opposite of supernatural, as when we say that there is a "perfectly natural" explanation for a strange phenomenon. On other occasions "natural" is the opposite of "artificial" ("her makeup looked so *natural*") or "unusual" ("it's unnaturally rainy for this time of year") or "cultural" (as in "nature versus nurture"). In still other contexts "natural" refers to how things *should* be, as when we describe abnormal behavior as "unnatural" (an "unnatural act"). But none of these quite captures the sense in which war is natural.

The naturalness of war lies in its role as an innate, biologically based potential: something that nature has built us to be capable of. Potentials are like coiled springs. They are events waiting to happen. For example, right now, as I type these words, there is a half-empty wine glass perched precariously near the edge of my desk. The glass is fragile, and one careless move of my arm will cause it to fall, strike the floor, and almost certainly shatter into myriad pieces. The glass has the *potential* to shatter—a potential that will only be realized under certain circumstances, such as striking a hard floor. The same broad principles come into play when we consider the behavior of living things. Every organism has a behavioral repertoire, a set of potentials that lie dormant until or unless they are triggered by something in their environment. Some of these are rigidly fixed and robotic, producing the same stereotyped behavior over and over again. A paradigmatic example is the behavior of the digger wasp (*Sphex ichneumoneus*). When the female of this species is ready to lay her eggs, she digs a burrow in the ground. She then catches a cricket, which she paralyzes with a single sting, and carries it back to the nursery. Arriving at the burrow with her prize, she deposits it at the entrance and goes inside, apparently to make sure that all is well. Once this little ritual has been performed, the wasp drags the cricket in, deposits her clutch of eggs, carefully seals the mouth of the burrow, and flies away. The comatose cricket will be a living larder for her young when they hatch. The wasp's behavior seems remarkably thoughtful and deliberate, but this is an illusion. Move the paralyzed cricket a few inches away from the burrow while she is inside making sure that all is well, and she will again drag it to the threshold and again go inside to make an inspection. The procedure is stereotyped, hard-wired, and automatic; *Sphex ichneumoneus* will repeat the entire routine in precisely the same way time after time if the cricket is moved whenever she enters the burrow.[12] This led the Indiana University cognitive scientist Douglas Hofstadter to coin the adjective "sphexish" to describe such rigidly programmed patterns of behavior.

An excellent example of complex, sphexish behavior is provided by the unseemly conduct of a bacterium called *Pseudomonas aeruginosa*, one of the many guests that live in the human gut, where they pay their keep by helping us digest our meals. Most of the bacteria that live in our intestines are benign organisms, and so is *Pseudomonas*—at least under ordinary circumstances. However, these minuscule organisms turn out to be fair-weather friends. Research suggests that they are equipped with sensors that can detect concentrations of interferon-gamma, a chemical released by the human immune system to help it coordinate its battles against invading microorganisms. When the bacteria get a whiff of interferon-gamma, their first response is to do a rough tally of their own numbers, and if the body count is favorable, they switch on genes that cause them to morph into dangerous parasites. Abandoning their placid existence, the tiny bacteria produce a biofilm, a sort of bulletproof vest that protects them from counterattacks by their host's immune system. Next, they pump deadly anthraxlike toxins into the intestinal wall, through which they then bore their way and enter the bloodstream. *Pseudomonas* is tough. Resistant to antibiotics, it can kill its victim in a couple of days.[13] Of course, these bacteria have no idea why they are destroying their host, or even that they *are* destroying their host: *Pseudomonas* is so primitive that it has no brain to think with. The bacteria undergo their Jekyll-to-Hyde transformation because they are designed by evolution to launch an attack when they detect concentrations of interferon-gamma. The entire operation is mindless and automatic. Special mechanisms in the bacterium "interpret" the presence of the chemical as a sign that its host's immune system is compromised, and therefore that this particular intestine is no longer a desirable neighborhood. So, with guns blazing, *Pseudomonas* blasts its way out to find a new home.

Human behavior is far from sphexish. It is protean, variable, and often unpredictable, but the fact that much of our behavior is not automatic does not mean that it is unconstrained by biology. There are fundamental similarities between the attack behavior of *Pseudomonas aeruginosa* and the warlike behavior of *Homo sapiens*, although they lie at opposite ends of this spectrum. Neither behavior is inevitable, and both occur only in response to precipitating conditions. For the bacterium, the conditions are very simple: there has to be a large enough population of them and a great enough concentration of interferon-gamma. For human beings, the conditions are much more difficult to specify. In later chapters, I will describe what it is that releases our violent, warlike tendencies, but there is much to consider before we get to that point. My first task is to make the case that war is a normal feature of human life, which I will do in the next two chapters.

3

OUR OWN WORST ENEMY

> For though it were granted us by divine indulgence to be exempt from all that can be harmful to us from without, yet the perverseness of our folly is so bent, that we should never cease hammering out of our own hearts, as it were out of a flint, the seeds and sparkles of new misery to ourselves, till all were in a blaze again.
>
> —JOHN MILTON, *THE DOCTRINE AND DISCIPLINE OF DIVORCE*

CENTURIES AGO, WE PICTURED OUR HOME, the earth, as lying at the hub of the cosmic wheel, with the sun, moon, stars, and planets, indeed the entire universe, revolving around us. Not content with being the center, around which the entire cosmos turns, we depicted ourselves as made in the image of the Almighty, who set us apart from the animal kingdom. Even today, most people regard their own species as the highest form of life.

William Shakespeare, writing around the year 1600, expressed a version of this lofty estimate of humanity using the tragic, conflict-ridden Hamlet as his mouthpiece.

> What a piece of work is man! How noble in reason! How infinite in faculties! In form and moving, how express and admirable! In action how like an angel! In apprehension,

how like a god! The beauty of the world! The paragon of animals![1]

We human beings possess many characteristics that set us apart from all the other denizens of Planet Earth. Some of these are quite astonishing. We possess an amazing and endlessly creative ability to use language, to think abstractly and harness the power of science to analyze the deep structure of the physical world from the tiniest quark to the remotest galaxies. Our mental powers are so formidable that we can chart the life cycle of the universe from the first yactoseconds of the Big Bang to its eventual entropic death many billions of years in the future. Thanks to our amazing mental powers we can reconstruct the trajectory of life on earth from the first replicating molecules, through a museum of now-extinct creatures right up to the panoply of living things that grow, walk, swim, fly, and slither over the earth's surface today. We alone are aware of the inevitability of death and create awesome works of art, magnificent religions, and profound philosophical systems to buffer us against the icy touch of mortality. And last but not least, we are the proud possessors of an unparalleled aptitude for cooperation, imitation, and collective action that enables us to produce and reproduce culture.

As uplifting as this picture undoubtedly is, it is incomplete. Our noble achievements are only half the picture. They exist side by side with an array of less appealing characteristics, the most disturbing of which is our propensity for committing acts of extreme cruelty and violence against our own species. Of all the residents of Planet Earth, we are the only animal that wages war (although, as we shall see, our relatives the chimpanzees do something very similar).

Nature imposes severe demands on her creations. Over 99 percent of the species that have ever lived are now extinct. Many of these extinctions were caused by climate change as, over its four-and-a-half billion-year history, the earth heated up and cooled down, was inundated with water and dried out, was exposed to the harshness of the sun, and languished under thick blankets of cloud and dust. But climate change is not the sole assassin. Species can also be driven to extinction by an unequal contest for limited resources. In the cut-throat world of nature, organisms that are outclassed in the struggle for survival—that are not sufficiently flexible to work out solutions to life's problems quickly or efficiently enough—are doomed. For example, the disappearance of most of the world's megafauna, large animals like the mammoth, the wooly rhinoceros, and the amazing ten-foot-tall elephant bird of Madagascar, coincided with the spread of *Homo sapiens* across the globe, and the evidence suggests that we hunted them out of existence.[2] *Homo sapiens* has performed magnificently in the struggle for survival, thanks mainly to the high intelligence that has made us so amazingly adaptable. In the five million years since our earliest ancestors opened their apelike eyes to the African dawn we have been able to handle every challenge that nature has thrown at us. When we were confronted with the plummeting temperatures of the last Ice Age we survived by donning thick animal pelts and warming ourselves at the fires that we had learned to build. Over the millennia, we learned to fight disease with herbs and hygiene and preserve life by caring for our sick and elderly. Even though early human beings were relatively small and weak, they learned how to manufacture weapons and put them to use in cooperative hunting techniques that were so deadly that they managed to defeat the large carnivores that had previously terrorized them.[3] Our path through time has been a sequence of triumphs over adversity, an evolutionary rags-to-riches story from which we emerged as seemingly invincible masters of the planet. However, the human story carries an ironic sting: the more successful human beings became at defending themselves against nonhuman predators, the more dangerous they became to one another. Eventually, our ancestors were forced to contend with a killer that was vastly more formidable, more diabolically intelligent and ruthless than the lions and hyenas: *they had to defend themselves against one another.* At the dawn of prehistory, they fought with hands and teeth, and later with sticks and stones. Farther down the road, they learned the art of chipping away at chunks of stone to craft flint blades and, hafting them to wood, created an array of knives, axes, arrows, and spears. Later still, they discovered how to fashion death-dealing instruments from bronze and eventually mastered the use of iron. The domestication of the horse led to mounted warfare and to large, highly mobile armies. The discovery of gunpowder and the invention of firearms unleashed a cascade of advances in the technology of killing. Throughout our history, each advance made us more lethal. Today we possess weapons of unprecedented destructive force capable of annihilating every person on the planet.

TWO MOMENTOUS STEPS IN THE HISTORY OF SLAUGHTER

The frequency of armed conflict across the globe and throughout history is staggering. I mentioned earlier that around 87 million people have been killed in wars over the last century. If we add the victims of democide we get, at a conservative estimate, around 170 million deaths in the twentieth century alone. That comes to an average of 1.7 million a year, 4,630 a day, 193 an hour and 3 a minute. In the 1990s alone, around 2 million were killed in Afghanistan, 1.5 million in Sudan, almost 1 million in Rwanda, 500,000 in Angola, 250,000 in Bosnia and the same in Burundi, 200,000 in Guatemala, 150,000 in Liberia, and 75,000 in Algeria, plus the cumulative toll of deaths from smaller conflicts. Looking at forty-one modern nation-states between 1800 and 1945, we find that they average 1.4 wars per generation and 18.5 years of war per generation. Russia and the United Kingdom have been the worst offenders, with 3.6 and 5.9 wars per generation (and 49.3 and 48.3 years of armed conflict per century respectively), followed by France, Spain, Turkey, and Italy.[4]

Ever since the philosopher Jean-Jacques Rousseau wrote in his 1754 *Discourse on the Origin of Inequality Among Men* that the first human beings were "noble savages" who lived their lives "wandering up and down the forests . . . an equal stranger to war and to all ties, neither standing in need of his fellow-creatures nor having any desire to hurt them" there has been a tendency to deny that war is part of human nature.[5] According to this optimistic scenario, war only came about when the development of agriculture turned peaceful nomadic foragers into tillers of the soil. The evidence, however, presents a much more somber picture.

There is no archeological evidence that throws light on group violence during the earliest evolution of the hominid line. In all probability, our earliest ancestors behaved like chimpanzees. Small groups of males patrolled their territory and made occasional deadly forays into their neighbors' range. During this period, small, more recognizably human, foraging bands roamed a landscape rich in wild game. Over time, these groups perfected cooperative hunting and developed weapons and tools for bringing down, skinning, and butchering large game. They crafted stone ax heads and spear points, and at least 400,000 years ago they were producing graceful, eight-foot-long wooden spears resembling modern javelins. The bow made its debut during this remote epoch, and played a vital role in hunting and warfare for thousands of years to come. At some point, these warlike practices probably began to evolve into organized, armed raids.

Prehistoric people turned their weapons against human quarry. The exquisite, strangely evocative murals left on cave walls by late Paleolithic artists mainly depict large mammals such as deer, wild horses, cattle, wooly rhinoceroses, and ubiquitous rumbling herds of bison, painted in rusty shades of iron oxide paste. However, a few portray human beings, and some of these show human beings killing one another. The caves at Pech Merle and Cougnac, in southwestern France, contain portraits of bodies impaled, pin-cushion style, with multiple arrows, and one man is shown standing with seven of them protruding from his body. The artwork at Morella la Vella in Castellón, Spain, portrays a line of archers and a supine figure pierced with ten arrows, as well as a battle between two groups of archers. Similar examples dating back to the Old Stone Age have been found in Italy's Paglicci Cave, as well as Combrel, Gourdan, and Sous-Grand-Lac in France. Rock paintings in Arnhem Land, in northern Australia, show groups of men locked fiercely in battle with spears and boomerangs hurtling through the air overhead.[6]

The emergence of organized raiding, modeled on cooperative

hunting and the use of weapons designed to fell game, was the first momentous step in the evolution of warfare. We do not know for certain, but there is evidence to suggest that these may have been cannibalistic hunts. Archeologists have found human bones scarred with the distinctive cut and scrape marks that are produced when an animal is butchered, and the smashed skulls and thigh bones found at some prehistoric sites may be evidence of our ancestors' taste for human brains and marrow.[7]

Some human remains from prehistory show unmistakable signs of violent assault. A Neanderthal man who lived in a cave in Shanidar, Iraq, was stabbed in the side in what was probably a face-to-face confrontation; a young child was killed 30,000 years ago in what is now Grimaldi, Italy, when a spear pierced his tiny body and lodged deep in his spine; an Egyptian man met his death 20,000 years ago when two arrows entered his abdomen and a third lodged in his upper arm. The most impressive of these remains is a Stone Age cemetery at Jebel Sahaba on the Egyptian-Sudanese border dating from around eleven thousand years before Christ. Nearly half of the skeletons discovered there belonged to men, women, and children who died violent deaths, with flint arrowheads and barbs found in or among their bones. Many have multiple wounds—sometimes up to twenty. The children were killed by arrows shot into their necks at close range, a sign that they were executed, perhaps after their mothers were raped and their fathers killed in battle.[8]

As the archeological record rolls on, we find more profuse evidence of mass killing. Tucked away among the verdant, gently rolling hills of Talheim, Germany, archaeologists discovered a grave containing what was left of the bodies of eighteen adults and sixteen children, all of whom were killed by blows to the head and tossed into a pit. Further south, at Ofnet Cave near Nördlingen, Bavaria, archeologists found 34 decapitated skulls of men, women, and children who were killed by blows to the head. The heads may have been taken as battle trophies (see chapter 10). At Roaix, in the south of France, they unearthed a prehistoric mass burial site containing around 100 skeletons, many of which have arrowheads embedded in them. Archaeologists digging at Crow Creek, South Dakota, found a grave containing the remains of over 500 adults and children who were scalped, mutilated, and killed about a century before Columbus arrived in the Americas. The massacre wiped out more than half of the settlement's original population, and at Cowboy Wash, Colorado, they found fossilized human feces containing traces of human muscle protein.[9] Many similar examples litter the literature on prehistoric archaeology. These are just a few examples of the many finds that testify to the existence of war in prehistory.[10]

The second giant step in the evolution of war took place during a time of rapid cultural change that archeologists call the Neolithic revolution. The Neolithic period, or New Stone Age, began around 8000 B.C., when nomadic tribes started to build permanent settlements in the Fertile Crescent, the rich swath of land that sweeps from present-day Syria east and south to what is now Iraq. These developments dramatically reconfigured human culture, which then radiated outward in successive waves of conquest, migration, and trade. The settlers exploited the wild cereals that grew in the region and soon learned to domesticate oats, wheat, and barley and to support themselves by cultivating the soil. As population swelled, and the once plentiful wild game dwindled, the settlers began to domesticate sheep, goats, cattle, and pigs, which gave them a steady source of milk and meat.

For the first time, groups were permanently tied to specific tracts of land upon which they depended for their survival. The new way of life provided food to support large, concentrated populations, and larger populations made room for an explosion of technology and symbolic culture, which culminated in the invention of writing.[11] However, many tribes held fast to the old, nomadic way of life. They spent their lives driving flocks of goats and sheep across the arid, marginal areas while the agriculturists put down roots in the most fertile areas, the biblical "land flowing with milk and honey." These pastoral nomads posed an ongoing threat to the jealously guarded resources of settled populations. They were, as military-intelligence analyst and historian Robert O'Connell vividly puts it, "wandering masses of mouths, eating their way across the landscape."[12] Over the centuries, the desert herdsmen perfected skills that came in handy against human enemies. "They knew how to break a flock up into manageable sections," writes military historian John Keegan, "how to cut off a line of retreat by circling to a flank, how to compress scattered beasts into a compact mass, how to isolate flock leaders, how to dominate superior numbers by threat and menace, how to kill the chosen few while leaving the mass inert and subject to control."[13] They knew how to kill quickly, coldly, and efficiently.

It may have been this ongoing threat that pushed early agriculturists to encircle their towns by massive protective walls, trenches, and earthworks. Whatever the reason, when the first cities sprang up approximately ten thousand years ago, their builders made sure that they were well protected. The ancient city of Jericho was secured against attack by earthwork embankments and protective walls twelve feet high and over six feet thick constructed from over twelve thousand tons of limestone. All of this was crowned by a lookout tower twenty-five feet tall. This may sound paltry by today's standards, or even by the standards of the great civilizations of Egypt and Mesopotamia that blossomed five thousand years later, but Jericho was an unprecedented feat of prehistoric engineering, a display of technological prowess that would have struck fear into the hearts of anyone audacious enough to consider attacking it (these facts give the biblical story of Joshua's sacking of Jericho its force, which is unfortunately lost on most modern readers). If the mighty walls of Jericho were built to "protect a horn of plenty, surrounded by a dry and hungry world," and the incredible labor involved in building them must have been proportionate to the severity of the threat. The design of Çatal Hüyük, a Neolithic city the ruins of which lie in what is now south-central Turkey, convey a similar impression. The architects of Çatal Hüyük, protected its 5,000 citizens in a different way. All of the buildings were interconnected, like a giant beehive, with sheer, windowless walls facing the outside, apparently to serve as a fortification against a world swarming with hungry enemies.[14]

The appearance of large settlements prepared the ground for a new cultural phenomenon. For the first time, human beings embarked on wars of conquest. The ancient agricultural societies were almost entirely dependent on the success of the annual harvest. They had virtually no buffer against crop failure, and drought, disease, or a plague of locusts could spell disaster, resulting in hunger, socioeconomic collapse, and death. A spate of successful harvests also had its problems, as this was accompanied by population growth, and consequent increase in demand for resources to feed hungry mouths. The solution to both problems was territorial expansion.[15] The hunger for territory inspired new developments in weapons technology. For the first time, there was industry devoted entirely to the manufacture of weapons specifically for killing human beings—weapons like the mace, the sling, the dagger, and the composite bow—the original weapons of mass destruction.[16]

It was at this fateful moment that humanity stepped across the threshold of the military horizon and "true" warfare entered the stage of history. As the new weapons proliferated, they were met with ever more imposing defensive fortifications in an escalating arms race that spurred rapid technological and architectural innovation.

The most evocative testimony to our legacy of violence is a lone mummified corpse found in 1991 by two hikers embedded in a melting glacier in Austria's Ötztal Alps. "Ötzi the Iceman," as he came to be called, was astonishingly well preserved. Dressed in a grass cape, goatskin leggings, and coat, he was also armed to the teeth with a stone-tipped knife, a copper battle-ax, a bow, and a quiver of arrows. Radiocarbon dating showed that the body was around five thousand years old. When scientists examined Ötzi's body, they discovered cuts and bruises on his abdomen, and DNA analysis revealed that the traces of blood that stained his weapons and clothing came from four different people. The examination also disclosed the cause of the iceman's death: a flint arrowhead buried deep in his shoulder. Ötzi was killed by another human being. Perhaps he was ambushed by a raiding party while out hunting and maybe the spatters of blood on his clothing are the remains of a deadly clash in which he took down three of his attackers before finally succumbing to the agony of his rapidly hemorrhaging shoulder. We will never know exactly what happened to Ötzi, but it is clear that his death was neither an accident nor an execution. It was the outcome of a violent struggle.

The period between the first permanent settlements in the Fertile Crescent and the invention of writing in Mesopotamia circa 3000 B.C. was a time of escalating violence. Fortified towns sprang up like mushrooms. Archeological discoveries of broken, mutilated corpses beneath the charred remains of what were once proud cities tell a tale of furious and incessant battle. Craftsmen turned their talents to fashioning deadly weapons made from bronze, and entire civilizations are snuffed out in wave upon wave of imperial expansion as the newly minted kingdoms flexed their military muscle. As populations swelled and weapons technology leaped ahead, fortifications expanded proportionately. By the third millennium B.C. the Mesopotamian city of Uruk could boast of defenses that dwarfed those of the once mighty Jericho—its huge walls were over six miles long and studded with nine hundred towers.

The early writings from Egypt, Sumer, Greece, Rome, India, and Mesoamerica record slaughter on a grand scale.[17] Of these, the Assyrian accounts were perhaps the most graphic. The triumphant King Sennacherib had no reservations about boasting of how ruthlessly he treated his enemies, the Elamites.

> I cut their throats like sheep. . . . My prancing steeds . . . plunged into their welling blood as into a river; the wheels of my battle chariots were bespattered with blood and filth. I filled the plain with the corpses of their warriors. . . . As to the sheikhs of the Chaldeans; panic from my onslaught overwhelmed them like a demon. They abandoned their tents and fled for their lives, crushing the corpses of their troops as they went. . . . [In their terror] they passed scalding urine and voided their excrement in their chariots.[18]

There are also ancient accounts of war from the perspective of the defeated, the most moving of which are the Mesopotamian songs of lamentation, which describe the horrors of war in terms that are as applicable today as they were four thousand years ago. When the Sumerian city of Ur fell to the Elamites in 1950 B.C., an anonymous poet sang:

> Then the storm was removed from the city, that city reduced to ruin mounds. . . . Its people littered its outskirts just as if they might have been broken potsherds. Breaches had been made in its walls—the people groan. On its lofty city-gates where walks had been taken, corpses were piled. On its boulevards where festivals had been held, heads lay scattered. In all its streets where walks had been taken, corpses were piled. In its places where the dances of the Land had taken place, people were stacked in heaps. They made the blood of the Land flow down the wadis like copper or tin. Its corpses, like fat left in the sun, melted away of themselves.[19]

Homer's *Iliad*, the greatest literary monument of the Hellenic world, contains descriptions of carnage that are so graphic that the person who composed them must have had direct battlefield experience. In a spirit far removed from the despair of Sumerian lamentations, the *Iliad* describes a world in which warriors revel in the glory of bloodshed, hacking one another to bits on the plains of Troy.

> Achilles then went up to Mulius and struck him on the ear with a spear, and the bronze spear-head came right out at the other ear. He also struck Echeclus son of Agenor on the head with his sword, which became warm with the blood, while death and stern fate closed the eyes of Echeclus. Next in order the bronze point of his spear wounded Deucalion in the fore-arm where the sinews of the elbow are united, whereon he waited Achilles' onset with his arm hanging down and death staring him in the face. Achilles cut his head off with a blow from his sword

> and flung it helmet and all away from him, and the marrow came oozing out of his backbone as he lay. He then went in pursuit of Rhigmus, noble son of Peires, who had come from fertile Thrace, and struck him through the middle with a spear which fixed itself in his belly, so that he fell headlong from his chariot. He also speared Areithous squire to Rhigmus in the back as he was turning his horses in flight, and thrust him from his chariot, while the horses were struck with panic.[20]

Peruse the ancient myths and chronicles, and you will discover story after story of slaughter and destruction. Look at the paintings, the sculptures, the texts, inscriptions, and bas-reliefs that have survived the depredations of time and you will find depiction after depiction of gods, kings, and armies mercilessly obliterating their enemies. In earliest prehistory, the sparseness of human population put a natural brake on the scope our ancestors' destructiveness. But when agriculture made large concentrations of people possible, and urban centers began to turn nomadic tribesmen into sedentary tillers of the soil, the deadliness of war increased proportionally. There were more people to kill and increasingly rich resources to seize from them. By the fifth century B.C., empires had become so large, and populations had become so dense, that battles could be massive engagements. According to the Greek historian Herodotus, his countrymen wiped out over 200,000 Persians on a single summer day in 479 B.C. at the Battle of Plataea. Almost two hundred years later, in present-day China, the army of the kingdom of Qin slaughtered almost a quarter of a million of the combined forces of Han and Wei.[21]

The transition from the world of the ancient empires to our own was a relatively small step compared to the immense social transformation that took humanity from the Stone Age to the epoch of recorded history. An ancient Assyrian soldier would be astonished at our sophisticated military technology, but he would be no stranger to our concepts of military engagement and strategy, which have changed relatively little in the past two or three thousand years. However, a man stepping out of the Old Stone Age into the armor of an Assyrian infantryman, or the uniform of a contemporary soldier, would be stupefied. This would not be because of any lack of intelligence or aggression (our prehistoric ancestors must have possessed both in abundance), but because the social framework of human life, and the scale of military activity, had changed so drastically.

GENOCIDE

War is often genocidal, aimed at destroying entire peoples rather than simply defeating an enemy. This has been so from the time that war began. But genocide is not a relic of the past, safely confined to the bloodless, two-dimensional pages of history books. As you read this sentence a genocide is occurring somewhere in the world, and others are being planned (as I write this sentence, the long-standing genocide in the Darfur region of Sudan is claiming 500 souls every day). The cumulative death toll of the major genocides of the last century alone exceeds 30,000,000 lives (see appendix).

Although war is supposedly at odds with the core values of the Judeo-Christian tradition, the Holy Bible recounts many episodes of mass destruction of human lives implemented by God or undertaken at his behest. Chapter 7 of Genesis reveals that the Almighty "opened the windows of heaven" to send a flood that drowned the entire human race except for Noah and his family. Later, in chapter 19, he rains down burning sulfur on the cities of Sodom and Gomorrah, incinerating all the residents. (Sulfur melts as it burns. In effect, the Almighty napalmed every man, woman, and child in the two cities, subjecting them to an extremely agonizing death.) In Exodus, God first hardens Pharaoh's heart so that he stubbornly refuses to release the Israelites from captivity and then brutally punishes the Egyptians by sending ten plagues to torment them, culminating in the death of all firstborn Egyptian men, women, and children. In the same book, God promises the Children of Israel, "I will send my terror before you . . . and will throw into confusion all the people." This promise is made good in the story of conquest recounted in the books of Deuteronomy and Joshua, where we are treated to an account of how Moses and Joshua embarked on a campaign of Holy War (in Hebrew, *herem*) to exterminate the inhabitants of Canaan. In Deuteronomy God informs Moses: "In the cities of the nations the LORD your God is giving you as an inheritance, do not leave alive anything that breathes. Completely destroy them—the Hittites, Amorites, Canaanites, Perizzites, Hivites, and Jebusites—as the LORD your God has commanded you."[22] The scriptures provide a blow-by-blow narrative of Joshua's campaign, describing how his troops slaughtered every human being in their path as they cut a bloody path across the Fertile Crescent: "He left no survivors. He totally destroyed all who breathed, just as the LORD, the God of Israel, had commanded."[23]

The close association between faith and genocide exemplified by the Book of Joshua is no accident. Throughout history the seeds of genocide and other atrocities have often been watered by religious zeal—principally, but not exclusively, by the missionary and monotheistic religions of Christianity and Islam. Of course, the persecution of European Jews, which reached its apogee in the Holocaust, is the best-known example. The Nazis built the Final Solution on Christian foundations going back almost two thousand years.[24] Practically from its inception, Christianity targeted the Jews, and the long list of Christian writers whose teachings incited violence against them includes a bevy of Catholic saints and even Martin Luther.[25] The program of anti-Jewish terrorism nurtured in the womb of Christianity was applied in a series of expulsions, pogroms, inquisitions, and crusades spanning many centuries.

The gospel story has Jesus announce, "Do not suppose that I have come to bring peace to the earth. I did not come to bring peace, but a sword."[26] His heirs certainly fulfilled this prophecy. Although Christians were initially forbidden to use violence, even in self-defense, their aggressiveness found other, spiritually sanctioned, outlets. Tertullian, who lived in the second and third centuries A.D., calmed his Christian brethren who were disconsolate over being forbidden to witness the gory gladiatorial contests in the Hippodrome by assuring them they would be able to witness even greater tortures after death, gazing down from heaven to observe enemies of the faith writhing in eternal torment.[27] Later, the conversion of Europe was accomplished in large measure under the shadow of the threat of death. During the Saxon wars of the eighth century the emperor Charlemagne converted the pagan tribes of Europe at sword point, and reputedly slaughtered 4,500 renegade Saxon noblemen at the bloodbath of Verden as punishment for reverting to their original religious practices (and a disincentive for any other ex-pagans contemplating doing the same).[28] The suppression of heresy was accomplished by similar means. By the ninth century, the popes began to declare that butchering unbelievers was good for the souls of warriors. Every life taken in the service of the Church was supposed to expiate a bit of the knight's backlog of unexcused sin.

Most educated people have at least a nodding acquaintance with the Crusades and the terrible wars between Catholics and Protestants during the Reformation, but fewer are aware of the ruthless persecution of heretical sects within Christianity itself. The Albigensian Crusade was the most savage of these. In 1209 Pope Innocent III organized a holy war against the Cathars, a large Christian group based in the Languedoc region of southern France. His war of extermination lasted for twenty years. The first engagement, in the town of Béziers, set the tone of the entire campaign. On a sultry July afternoon the papal army marched through the city gates under the leadership of Abbot Arnaud Amaury, head of the Cistercian Order and papal legate. The entire population of between ten and twenty thousand men, women, and children were cut down that day. "All were killed," wrote a contemporary chronicler, "even those who took refuge in the church. Nothing could save them. Neither crucifix nor altar." Only two hundred or so of these people were Cathars. The others were killed because they refused to surrender the Cathars to the pope's genocidal army.[29] Three and a half centuries later, Europe convulsed with slaughter as Catholics fought Protestants in the name of the One True God. It was during this time that Pope Gregory III declared that virtually the entire population of the Netherlands—some three million souls—were heretics. King Phillip of Spain ordered the Duke of Alba, a fanatical Catholic, to execute them, a task that he pursued with gusto until he was recalled to Spain six years later. At roughly the same time, untold thousands of French Protestants were killed in the weeks following the infamous St. Bartholomew's Day Massacre of 1572. Pope Gregory was so delighted with the outcome that he commissioned a mural to commemorate the event, which remains in the Vatican to this day.[30]

A less well known but far more momentous example of faith-based brutality was the Taiping rebellion in mid-nineteenth-century China. The rebellion was led by a peasant named Hung Hsiu-ch'üan who, after obtaining religious instruction from a Southern Baptist minister, came to believe that he was Christ's younger brother and that it was his mission to eradicate demons and demon worship from the world. Hung's gospel was attractive to many poor peasants, and he soon recruited an army of around a million famine-stricken followers who began to conquer territory in 1851. Hung declared the establishment of a Christian "Kingdom of Heavenly Peace" with himself as "Heavenly King." The era of Heavenly Peace ("Taiping") had begun. Hung banned the study of traditional Confucian texts and replaced it with mandatory biblical instruction. He also instituted a number of social reforms, including the abolition of opium, tobacco, alcohol, prostitution, slavery, and foot-binding. He also established equality for women and tried to eliminate private property. Like today's jihadists, Hung's troops burned with such religious ardor that they did not fear death and fought against the imperial army with astonishing discipline and horrific brutality. They looted, plundered, and slaughtered the inhabitants of entire cities who opposed their teachings. They committed atrocities and burned Buddhist and Taoist temples to the ground. The Taiping gained control of large swaths of south and central China by the time of their final defeat in 1864. Hung's religious movement had initiated the bloodiest civil war in history, which, when the final curtain fell on the Era of Heavenly Peace, had taken 20 to 30 million lives.[31]

Christianity has no monopoly on bloodshed. "O Prophet!" states the Koran, "urge the believers to war; if there are twenty patient ones of you they shall overcome two hundred, and if there are a hundred of you they shall overcome a thousand of those who disbelieve, because they are a people who do not understand."[32] Looking south to Africa, and east toward Asia, Islamic armies did precisely that. Early in Muhammad's war on the ungodly, he expelled two of the three Jewish tribes that lived in the city of Medina and beheaded all eight hundred men of the remaining Banu Kuraiza tribe in the city marketplace. He spared the lives of the women and children and sold them into slavery, all at the behest, it is said, of the angel Gabriel.[33] The Islamic conquest of the Indian subcontinent was one of the most ghastly episodes in Asian history; according to one contemporary account, "The blood of the infidels flowed so copiously that the stream was discolored, notwithstanding its purity, and people were unable to drink it. . . . Praise be to Allah for the honor he bestows on Islam and Muslims."[34]

Nearer to the modern era, Europeans invoked the biblical narrative of Joshua's conquest of the Promised Land to justify the expansion of European power in the New World. The destruction of the pre-Columbian cultures of North and South America is the most extensive genocide that the world has ever known. Soon after Christopher Columbus planted the Spanish flag on the shore of Hispaniola, he began the destruction of the indigenous Taino people. Men, women, and children were killed and tortured, roasted on spits, and hacked to pieces to feed the Spaniards' dogs, and in the space of fifty years approximately 3 million people were reduced to around two hundred.[35] Estimates of the original native population of the Americas vary dramatically, and it is unclear what proportion of the millions of deaths was caused by contagious diseases unwittingly brought by the European settlers, but it is certain that violent aggression accounted for millions, if not tens of millions, of fatalities.[36] One of the most infamous of these massacres in the nineteenth century was the 1864 cavalry attack led by the "fighting parson" John Chivington on a village of approximately five hundred unarmed Southern Cheyenne and Arapaho at Sand Creek, Colorado. Almost two hundred—mostly old men, women, and children—were massacred. Witnesses testified before Congress that "bodies were mutilated in the most horrible manner—men, women and children's privates cut out . . . I heard one man say that he had cut out a woman's private parts and had them for exhibition on a stick." Another said that the Indians "were scalped, their brains knocked out; the men used their knives, ripped open women, clubbed little children. . . ."[37] Similar events took place all over the world, as colonial expansion displaced indigenous people. The inhabitants of Tasmania were hunted to extinction by their "civilized" British invaders, and the Dutch meted out the same treatment to the San people in South Africa. Simultaneously, the transatlantic slave trade condemned millions of human beings to the holds of ships bound for the New World, vast numbers of whom did not survive the journey, while those who lived were bought and sold by plantation owners whose vast prosperity was built on the backs of enslaved Africans. At the same time the sale and purchase of human flesh was a thriving business in slave markets from Morocco to China.

This summary account includes only a minuscule fraction of the terror and bloodshed that human beings have inflicted on one another over the centuries. The subject is so vast, the episodes so frequent, and the details so horrendous that it defies description. The history of humanity is, to a very great extent, a history of violence.

IGNOBLE SAVAGES

As we have seen, the idea of universal prehistoric peace is a myth. "If anything," writes anthropologist Lawrence J. Keeley, "peace was a scarcer commodity for members of bands, tribes and chiefdoms than for the average citizen of a civilized state."[38] Anthropological studies also reveal that war is common in hunter-gatherer and tribal societies. However, a word of caution is in order here. Compiling statistics on the incidence of war in primitive societies is a tricky business. The results of any such investigation will turn on precisely how the researcher defines "war." Is war distinguished from raiding? From blood feuds or revenge killings? The way that one answers to these and similar questions will have a decisive impact on the outcome of any study of indigenous warfare.[39]

In one classic study, Yale University anthropologist Carol Ember found that 64 percent of hunter-gatherer societies for whom records exist engage in warfare at least once every two years, 26 percent are at war somewhat less frequently.[40] Just 10 percent are rated as peaceful. In a survey of 157 Native American societies, only 13 percent did not engage in raids more than once a year, and only seven, a paltry 4.5 percent, turned out to be completely peaceful, and all of these were groups living in conditions of extreme geographical isolation and low population density.[41] These conflicts are serious, and sometimes genocidal. "Rarely were prisoners taken in tribal-level warfare, except for women who were integrated into the victors' society," writes Harvard anthropologist Steven A. LeBlanc. "The goal is annihilation of all men, women and children, although men were the primary target."[42] One of many such examples in the ethnographic literature is the Kutchin, a subarctic Native American group. The Kutchin would encircle an enemy village and then converge on it to kill every male inhabitant except one, called "the Survivor," whose life was spared so someone would live to tell the tale, and spread the word, of the slaughter. Among the Yanomami, 37 percent of all males die in warfare, as do a staggering 59 percent of males of the Jivaro, who inhabit the rain forests of Ecuador and Peru. Comparable, if not quite so extreme, figures have been recorded for the Mae Enga and Dugum Dani of New Guinea (35 percent and 28 percent respectively) and the Murngin of Australia (28 percent).[43] The gentle hunter-gatherer societies of popular imagination turn out to be very rare, and those few that genuinely exist often possess some other rather unappealing features. Many are refugee groups depleted by genocide who barely manage to eke out an existence in geographical isolation from others and have extremely elevated homicide rates. Others live under the restraining yoke of a peace enforced by the modern nation-state of which they are a part.

Social scientists who study a culture for a relatively brief time risk drawing the false conclusion that war is entirely foreign to it. But who, looking only at today's Germans, would have an inkling that in living memory their forebears pursued an unprecedented campaign of genocide and global conquest. Would an anthropologist from Mars suspect that less than a century ago the great-grandparents of contemporary Belgians brutally killed and mutilated millions of human beings in the Congo? Over time, rampaging Vikings have morphed into placid Scandinavians, the crack legions of ancient Rome were the forefathers of today's notoriously unmilitary Italians, and the neutral Swiss are the descendants of swaggering Celtic warriors. Conversely, placid societies can also turn nasty. During the 1950s the gentle Semai of Malaya were caught up in hostilities between the British colonists and communist insurgents. Although traditionally opposed to violence, the Semai tribesmen, once they began to suffer casualties, showed a very different side of their character: "We killed, killed, killed . . . ," reported one Semai man. "We thought only of killing . . . truly we were drunk with blood."[44]

The more we learn about of ourselves and our history, the more we are confronted with our extraordinarily violent character. An obvious question raised by all of this is "Why?" I will begin to examine this vital question in the next chapter.

4

THE ORIGINS OF HUMAN NATURE

The earth, like a wife,
The sky, like the husband,
Would, in a behavior like that of a cat,
Eat their offspring.
I do not understand such a woman,
I do not understand such a spouse.
I do not understand.

—JALALUDIN RUMI, *DIVAN E-SHAMS*

OF ALL THE CREATURES that have ever dwelt upon the earth, we are the only ones that devote untold effort and material resources to the task of cruelly exterminating members of our kind. The figures are sobering. In 2004 the world spent one thousand thirty-five *billion* dollars (about $2.8 billion each day) on armed forces, whereas aid to developing countries amounted to a mere $78.6 billion—less than 8 percent of the expenditure on arms. We devote far more effort to attacking other human beings, and preparing to defend ourselves against anticipated attacks, than we do on improving the human lot. In a world where millions are hungry, illiterate, and deprived of the most rudimentary comforts of civilized life—a world where horrendous diseases still decimate the poor and where environmental degradation threatens to destroy civilization itself—gigantic resources continue to be poured into the gaping maw of the military. As we have seen, the strange partnership between humanity and mass destruction is not a distinctively modern phenomena. It haunts the farthest recesses of our history.

A succession of brilliant thinkers has grappled with this paradox. In ancient times, war was attributed to the whims and machinations of divine beings, as part of the Almighty's inscrutable plan for humanity, as a manifestation of metaphysical evil, or a perverted expression of the free will with which we are all supposedly endowed. As scientific knowledge progressed, neuroscience and psychology began to transform the study of the human mind into a scientific enterprise. Sigmund Freud was one figure from this era who struggled to understand why, despite its horror and barbarity, human beings wage war. Working in the politically volatile ambience of interwar Austria, Freud wrote, "Men are not gentle, friendly creatures wishing for love, who simply defend themselves if they are attacked, but that a powerful measure of desire for aggression has to be reckoned as part of their instinctual endowment." Putting flesh on the bare bones of this idea, he added:

> As a result their neighbor is for them not only a potential helper or sexual object, but also someone who tempts them to satisfy their aggressiveness on him, to exploit his capacity for work without compensation, to use him sexually without his consent, to seize his possessions, to humiliate him, to cause him pain, to torture and to kill him.[1]

Freud was confident that science alone can unlock the mysteries of human behavior and ultimately provide a satisfactory explanation for war. But he was also aware of the speculative character of his own psychological theories, and argued that a real science of human behavior must stand on secure biological foundations. "The deficiencies in our description [of the mind] would probably vanish," he wrote in 1920, "if we were already in a position to replace the psychological terms by physiological or chemical ones."[2] The scientific landscape has changed dramatically in the almost nine decades since Freud wrote these words. Sleek technologies for brain imaging now allow scientists to peer into living brains in real time, sophisticated methods of experimental psychology enable them to study mental processes under controlled laboratory conditions, and mathematical and computational techniques make it possible to precisely model complex patterns of human behavior. These investigative techniques are overwhelmingly superior to the primitive tools that were available to earlier generations. We have better theories, too. Advances in genetics, evolutionary biology, and paleontology enable contemporary researchers to explore difficult questions about human nature in the context of novel and highly informative scientific frameworks. We are fast transforming Freud's dream of a biological science of the mind into a reality.

EVOLUTION

To comprehend war, we need to understand the biological factors that molded us into what we are. Earlier writers sometimes mentioned biology in their discussions of war but usually did so mainly in ways that were vague and uninformative. For example, the anthropologist Maurice R. Davie remarked in 1929 in *The Evolution of War*, "Since war is so fundamental a phenomenon its explanation must be sought in the basic conditions of life."[3] Although evolutionary biology is *the* discipline that investigates the impact of the "basic conditions of life" on living things, and in spite of the book's title, Davie was not much interested the relationship between biological evolution and war.[4] During the era when he wrote, the scholarly turf was neatly divided between the natural sciences (like biology, chemistry, and physics) and the human sciences (like psychology, sociology, and anthropology). Most people believed that human nature somehow transcended the merely physical world and thought that the world of atoms and molecules, cells and tissues investigated by natural scientists in their laboratories had nothing to contribute to understanding the intricacies of the human psyche. To some extent, this barrier persists today. There is still a divide between the so-called social sciences and the natural sciences, but it is gradually going the way of the Berlin Wall. Several decades of research have demonstrated beyond any reasonable doubt that evolutionary science gives us a uniquely powerful grasp on the dynamics of human behavior and experience. Evolutionary biology has made deep inroads into psychology, anthropology, and economics, and there is even a growing rapprochement between biological thinking and the study of literature that is starting to displace the antiscientific postmodernist consensus. And why not? We are organisms with a long and eventful evolutionary history. Our brains were shaped the same way as the organs of every other organism that has existed in the four and a half billion years since life began on earth. Notwithstanding our unique characteristics, human behavior is just as amenable to biological explanation as is the behavior of any other creature. For we are a part of the great family of all living things, not some anomalous entity set apart from the rest of nature and to whom only special explanatory principles apply.

Evolutionary biology began with the work of Charles Darwin in the first half of the nineteenth century. The son of a physician, he was sent away to study medicine at the University of Edinburgh to follow in his father's footsteps. Darwin detested medical school, and dropped out. Moving back to England, he enrolled at Cambridge University with the intention of studying theology and becoming a country vicar. However, while in Cambridge, contact with leading British naturalists fanned his interest in the natural rather than the supernatural world, and when he was offered the opportunity to join a maritime voyage around the world, collecting specimens and making observations of flora and fauna, he jumped at the chance. While on this odyssey Darwin made the observations that led him to his theory of evolution (or "descent with modification," as he preferred to call it) that was presented to the world in *On the Origin of Species.* This epoch-making book set out the explanatory framework without which, in the words of the geneticist Theodosius Dobzansky "nothing in biology makes sense."[5] Before Darwin, most people assumed that God had originally created all plants and animals, because they seemed to be too precisely engineered to have come about through undirected material processes. "Wherever we see marks of contrivance," wrote the nineteenth-century biologist William Paley, "we are led for its cause to an intelligent author."[6] Darwin demonstrated that we can account for these "marks of contrivance" by a few simple, natural principles operating over huge expanses of time. The creationists were right to say that living things bear the hallmarks of design, but wrongly inferred that God was their designer. The true "designer" of all organisms, including human beings, is the process of evolution.

We tend to think of ourselves as more "highly evolved" than other animals. It seems obvious to many of us that we are superior to, say, the common tapeworm. Can a tapeworm design skyscrapers or write novels? Obviously not. Well, then! But let's change gears and consider things from a tapeworm's point of view. A thinking tapeworm would probably regard it as self-evident that its own species is more highly evolved than we are. "Human beings don't have segmented bodies, are unable to live in an intestine, and are limited to being either male *or* female, but never both at the same time." It would say, "They are obviously very primitive organisms." The tapeworm would be wrong, of course; every bit as wrong as the human chauvinist, for although it is true that the word "evolution" means "progress" or "improvement" in everyday speech, in biology it has no such forward-looking connotations. Evolution is just adaptive *change* across generations. Change is horizontal, not vertical. Nature has no hierarchy although, as hierarchy-obsessed social primates, we find it practically irresistible to impose one. The clash between the scientific and the everyday uses of the word "evolution" has sometimes led to basic misunderstandings. People who are not familiar with biology often understand evolution as a relentless march forward, from imperfectly evolved life forms to that pinnacle of creation, humankind. But this is a pre-Darwinian paradigm, redolent of the religious notion of the Great Chain of Being. Biology is more democratic. All living things, from the humble bacterium to the human being, are equally "highly" evolved insofar as each is very good at doing what nature designed it to do. Objectively speaking, there is no "higher" or "lower" in nature, only a dazzling diversity of forms and functions. Unlike the arcane theories of physics, evolution is very easy to grasp. It consists of three phases, which spiral through time in an endlessly recurring pattern of variation, selection, and reproduction. Nature brings forth a huge number of slightly different organisms, makes each one run the gauntlet of life, discards the failures, and retains the successful models for further tinkering. The kaleidoscopic variations that we observe in nature—the gradations and variations of color, form, strength, speed, and so on—are like guesses about which design will outperform its alternatives in the contest of life. Nature is a lottery in which only some organisms hold a winning ticket, and the prize is the opportunity to pass genes on to the next generation. Natural selection is endless trial and error, but there is no perfect design, no final answer to the question of how best to live, because the shifting demands of existence continually confront living things with unanticipated challenges. Evolution never rests.

Living things are born into a world where there is limited room at the table, a world of fierce competition for resources such as food, water, space, mates, and even life itself. This was first set out by the British economist Thomas Malthus, who wrote in his *Essay on Population* (1798) that "population when unchecked, increases in a geometrical ratio. Subsistence increases only in an arithmetical ratio."[7] This means that, left to their own devices, populations expand rapidly, but the resources needed to support them increase more slowly, if at all. Any group that reproduces without inhibition will sooner or later outstrip the resources upon which it depends, and as the population swells individuals have no alternative but to compete more and more intensely for dwindling resources. Those who can secure them will flourish, and those who cannot will die. In a brilliant flash of insight, Darwin realized that Malthus's theory could explain the process of evolutionary change, and scribbled in his notebook that "it at once struck me that under these circumstances favorable variations would tend to be preserved and unfavorable ones destroyed. The result would be the formation of a new species."[8] Some individuals, he reasoned, are born with characteristics that help them in the struggle for survival. These lucky winners in the lottery of life may be a bit more robust than their peers or a tad more able to adjust to shifting climatic conditions. They may have a pattern of coloration that camouflages them from the hungry eyes of predators, have a bitter taste, or be fast enough to outrun those that would otherwise devour them. Of course, there are many other possibilities. The point is that the bearers of these life-enhancing variations are likely to survive longer than their peers do, and will therefore have a greater chance to reproduce and transmit the precious traits to their offspring. The youngsters will, in turn, pass the trait on to *their* offspring, and so on down the generations. Given enough time, and a relatively constant environment, the new characteristic will suffuse through the entire population, eventually reaching a point of "fixation," when every member of the species possesses it. Darwin called this whole process "natural selection" or, more graphically, "selection by death."[9]

Biology teaches us that life is never free from the looming shadow of death. The processes that drive the engine of evolution are hideously cruel. As Darwin put it in a letter to the American botanist Asa Grey, "I cannot persuade myself that a beneficent and omnipotent God would have designedly created the *Ichneumonidae* with the express intention of their feeding within the living bodies of caterpillars, or that a cat should play with mice." The *Ichneumonidae* are parasitic wasps that have a gruesome method for resourcing their offspring. The pregnant wasp catches a caterpillar or some other insect and lays its eggs on or in the victim's body so that when the larvae hatch they can gorge themselves on their host's living flesh. When the eggs hatch, the larvae eat the caterpillar alive. (Some wasps have taken this grisly procedure a step further. *Cotesia congregata* deposits her eggs inside her victim's body along with a virus that knocks out its immune system, which would otherwise attack the eggs.)[10] If you are unmoved by the suffering of insects, think of the comparable plight of members of your own species. The natural world is full of parasites that nature has designed to exploit the cozy environment of the human body, sometimes with grotesque or lethal results. Science writer

Carl Zimmer gives examples in his hair-raising book *Parasite Rex* including:

> Guinea worms, two-foot-long creatures that escape their hosts by punching a blister through the leg and crawling out over the course of a few days. Then there are filarial worms that cause elephantiasis, which can make a scrotum swell up until it can fill a wheelbarrow. Then there are tapeworms: eyeless, mouthless creatures that live in the intestines, stretching as long as sixty feet, made up of thousands of segments, each with its own male and female sex organs. There are leaf-shaped flukes in the liver and blood. There are single-celled parasites that cause malaria, invading blood cells and exploding them with a fresh new generation hungry for cells of their own.[11]

Most animals spend the bulk of their short lives either trying to kill others or trying to avoid being killed by them. Some are death machines, flesh-and-blood terminators built to track, capture, and butcher others with chilling efficiency. Others, like the nightmarish creatures described by Zimmer, are specialists in the covert operations of invading and exploiting the bodies of their unsuspecting hosts. Still others, the gentle herbivores, are mere predator fodder. Most animals are intermediate links in the food chain, born to be consumed alive by a predator that will in time fall victim to the same fate. Looked at in this way, the cycle of life and death seems heartbreakingly futile. The nineteenth-century German philosopher Arthur Schopenhauer put this with characteristic bluntness. Nature is a "scene of tormented and agonized beings, who only continue to exist by devouring each other, in which, therefore, every ravenous beast is the living grave of thousands of others, and its self-maintenance is a chain of painful deaths."[12]

Natural selection is only one part of this story. There is also another process underpinning evolutionary change which Darwin called "sexual selection."[13] Living to a ripe old age is no guarantee of reproductive success; sexually reproducing organisms have got to be sufficiently alluring to the opposite sex to get a shot at making babies. Now, every species has its own standards of attractiveness. Male chimpanzees go ape over females with large pink swellings on their hindquarters, while female Thompson's gazelles lust after reckless bucks that advertise their vigor by dancing defiantly in front of prowling lions. Whatever the criterion, outstandingly desirable individuals have disproportionately plentiful sexual opportunities. Because these ultrasexy individuals have high-quality mates at their beck and call, they can reproduce far more profligately than their less appealing peers. As a result, their genes, including the ones responsible for their sexually desirable traits, proliferate as effectively as the genes for traits that confer survival. A short-lived individual with many sexual opportunities can be just as (or even more) successful at reproduction than a long-lived individual that rarely mates, or that mates with down-market partners.

Given enough time, incremental changes add up to major reconfigurations. Operating over eons, natural and sexual selection have the power to shape and reshape living things. Darwin made the case that these two simple processes are jointly responsible for the lush variety of life. However, the theory of evolution does more than tell us how the stupendous diversity of nature came about, it also explains why organisms have their particular characteristics. According to evolutionary theory, traits possessed by living things are usually *adaptations:* characteristics that were selected because they are solutions to problems. Adaptations "make a difference." They help a creature to find food, repel disease, regulate body temperature, digest nutrients, escape predators, or perform some other vital function. The human brain is an organ that was sculpted by evolution, and this implies that our psychological traits—the ways that we think, the emotions that we experience, our tastes and preferences—are also adaptations.

One of the most important features of adaptations is their context sensitivity. Because they create a "fit" between an organism and the specific environment in which it lives, a trait that is adaptive in one environment can prove disastrous in another. Consider a winter coat. It is advantageous to wear a warm coat if you are outdoors on a cold winter day, but if you then go indoors, to a well-heated room, the coat becomes uncomfortable. Now, imagine that the coat is attached to your skin, so that you cannot take it off. This is precisely the condition of the polar bear, which has a thick coat of fur and ample subcutaneous fat to help it deal with bitter arctic temperatures. However, if global warming continues, these adaptations will become hindrances. Faced with environmental change, species must transform themselves or die. When environments alter, and existing adaptations lose their value, selection gets to work eliminating and replacing them. However, the process of evolution operates within a vastly longer time frame than the lifetimes of individual organisms, and it is therefore common for formerly adaptive characteristics to remain in a population long after they have lost their utility. Have you ever wondered why normal adults living in urban environments like Manhattan are liable to be terrified of snakes and spiders, while being quite blasé about dangers like cars and cigarettes? The answer is simple and obvious. The primate ancestors of present-day urban populations evolved millions of years ago in the forests and grasslands of Africa, where poisonous snakes and spiders posed a genuine danger.[14] By the same token, there were no cigarettes or cars around during the Pleistocene and before. The fear of snakes and spiders is no longer adaptive for a significant proportion of the world's population, but natural selection has not had the time to eliminate them and install other, more practical, fears in their place.

There were two gaping holes in Darwin's theory, which were filled only after his death. Darwin had no satisfactory account of the biological mechanisms that transmit traits, from one generation to the next, nor could explain the means by which nature innovates, that is, how variations arise in the course of evolution. These were major shortcomings, and they severely undermined the cogency of evolutionary theory. The solution to both problems lay dormant in the work of a man named Gregor Johann Mendel, a Czech monk whose meticulous experiments in the monastery garden demonstrated that plants contain microscopic chunks of information, which Mendel called "hereditary elements," that are passed from parents to offspring in the act of reproduction. Mendel had no idea of what these hereditary elements were, or exactly how they did their work, but the precise patterns of inheritance revealed in his experiments convinced him that although invisible to the naked eye, they *must* exist. His research was ignored for almost half a century until the Dutch botanist Hugo de Vries discovered it in 1900. De Vries renamed the hereditary elements "pangenes" (Greek for "that which gives birth to all"), and five years later another botanist, a Dane named Wilhelm Johansson, shortened this to "gene." Before long, the new science of "genetics" was under way. By the 1930s scientists were at work combining the new science of genes with Darwin's theory of evolution to create what became known as the modern synthesis.

Genetics solved both of Darwin's problems in one stroke. It identified the mechanism of heredity and demonstrated that nature innovates by means of random genetic changes called "mutations." When a gene mutates, nature rolls the dice. Sometimes the change is beneficial or neutral, and the organism wins, and sometimes it loses, when the change turns out to be harmful. But Mother Nature shoots craps with loaded dice; random changes are far more likely to be harmful than beneficial. However, every once in a while a mutation occurs that improves on the original, and it is likely to be preserved by selection. Imagine, by way of an analogy, a text that is copied and recopied by scribes over the course of thousands of years. Sometimes the scribe commits an error by accidentally omitting, misspelling, or inserting a word. Errors usually degrade the text, turning sense into nonsense. If too many of them accumulate, the document is reduced to gibberish and gets discarded. However, occasionally—very occasionally—a mistake is made that *improves* on the original. When this rare, serendipitous event happens, the "mistake" is retained, and what started out as a glitch becomes an adaptive feature. As time rolls on, and more and more beneficial errors are retained, the character of the text is gradually transformed into something far removed from the original version. After many reproductions, and many tiny modifications, documents derived through separate lineages from their common ancestor end up being dramatically different. In much the same way, natural and sexual selection act on random genetic variations by copying fortuitous mutations into the text of life.

To understand war, we need to find its place in the grand saga of evolution. We must tease out its deep biological roots by examining the role of aggression in the lives of nonhuman species how this heritage was transformed as our primate ancestors evolved into human beings. It is to this task that we will now turn.

NONHUMAN AGGRESSION

Concepts of animal aggression have changed remarkably over the last sixty years. The Nobel Prize–winning zoologist Konrad Lorenz popularized the idea that nonhuman species almost never engage in lethal aggression against their own kind, and therefore that *Homo sapiens* is an anomaly, a gross exception to the elegant norms that guide the behavior of all other animals. Lorenz believed that when animals of the same species fight, the potential for deadly violence is held in check by aggression-inhibiting mechanisms. Rivals try to intimidate rather than destroy opponents by engaging in "tournaments"—ritualistic, stereotyped posturing that is more like a fencing match or a dancing contest than a life-and-death struggle. For example, when male rattlesnakes compete for the attention of females, they forego the use of their deadly fangs and have a wrestling match instead, from which the loser can slither away and live to fight another day. When the piranha, a little fish renowned for its razor-sharp teeth and voracious feeding habits, fights members of its own species it does not bite but instead whacks the opponent with its inoffensive tail. Male anoles, small lizards native to the southern United States, the West Indies, and Latin America, "fight" by extending a colorful throat pouch and performing rapid bobbing movements to intimidate one another. In all of these examples, the individual that delivers the most impressive performance wins the day.[15] Nonhuman species also have a repertoire of submission or appeasement behaviors that switch off the aggression of an actual or potential attacker: gestures like lowering the head, grooming, lying supine, or assuming the posture of a sexually receptive female.

It is easy to see why nature might favor this way of doing things. Uncontrolled aggression can be extremely costly to winners and losers alike. It seems much more sensible for weaker individuals to yield to the larger, stronger, or more ferocious conspecifics than for them to risk duking it out with them. Lorenz believed that aggressive displays sometimes escalated and got out of hand, causing an occasional fatality—a manslaughter rather than a murder. However, as the science of animal behavior progressed, it became clear that Lorenz was not right. Many animals, including primates, kill infants of their own species. Although it is rare in most species for adults to inflict lethal damage on other adults—and to that extent Lorenz was correct—it is not *universally* rare. In some species deadly violence is relatively common, and their "murder" rate exceeds ours by a wide margin.[16] In the overwhelming majority of cases where animals kill their own kind, they do so in one-on-one encounters. For example, when male pronghorn antelopes battle with one another for mating privileges, 12 percent of these matches end in death for one or both combatants. Up to 29 percent of adult male red deer deaths are a consequence of fighting over females.[17] In these cases, the deaths are accidental. Red deer stags do not set out to murder one another. The killing is "collateral damage," a consequence of tournaments designed to win the favor of prospective mates. This is strikingly different from human warfare, which is a collective social phenomenon rather than an individualistic one. So, in our search for nonhuman prototypes for war we will have to narrow our focus and zoom in on what biologists call "coalitionary aggression." The best-known nonhuman coalitionary aggressors are ants. Some species swarm in their thousands to exterminate nearby colonies. In the United States, the imported fire ant (*Solenopsis invicta*) wages perpetual genocidal war against the native woodland ant (*Pheidole dentata*). Some ants not only attack their neighbors but also enslave them. The slave-making species *Polyergus breviceps* sends out scouts to locate any colonies of other species within a radius of one hundred fifty yards or so. Having located one, the scouts lead an army of several thousand minuscule amazon warriors to murder the enemy queen and carry the pupae (infants) back home, where they are reared as slaves to forage, feed, clean the nest, and tend to the queen of their masters. It is doubtful that we can learn much about ourselves by studying ants, as they are very distant relatives. We need to look closer to home. Wolves are far more closely related to us than ants, and wolf packs attack and kill other wolves with extraordinary frequency. In one study of Alaskan wolf packs, 39 to 65 percent of adult deaths were brought about by attacks by other wolves. Coalitional violence has also been observed among lions, spotted hyenas, and cheetahs, although it is uncertain how frequently it occurs in these species.[18] Wolves, lions, hyenas, and cheetahs are all members of the order Carnivora. But human beings are members of the order Primates. Think of the primate order as a large, extended family. As we all know, family members often have family resemblances, and the more closely any two individuals are related, the more similar they are likely to be, both physically and psychologically. Members of the primate order also have traits in common, such as opposable thumbs, flat fingernails and toenails, small noses, keen eyesight, eyes located at the front of the head instead of at the sides, long childhoods with extended maternal care, large brains, and relatively few teeth. Because we are primates, and all primates have certain features in common, it is possible to learn about ourselves from our nonhuman cousins.

The family tree of the primate order forks into two large branches. One branch consists of weird-looking animals called the prosimians, which include lemurs, lorises, and tarsiers. The prosimians are our most distant primate relatives. The other branch of the family tree divides into the New World monkeys and the Old World primates. The branch containing the Old World primates bifurcates into the Old World monkeys and the "hominoids," a group that includes the gibbons and orangutans of southeast Asia, the gorillas and chimpanzees of Africa (including the bonobo, or "pygmy chimpanzee,") and *Homo sapiens*. Looking at the behavior of other primates should be much more illuminating than the study of carnivores like wolves and hyenas. In particular, we ought to concentrate on those primates that are our closest biological relatives: the chimpanzees. Chimpanzees and humans diverged from a common ancestor around five million years ago and share a whopping 98.77 percent of their genes. If coalitionary violence is in our genes, it is probably in their genes, too. Before zooming in on chimpanzee violence, though, I need to lay out some general characteristics of the social mammals that will help us understand human behavior.

INTOLERANCE

Some animals are loners, who prefer an isolated, hermitlike existence. Others are gregarious, preferring the hustle and bustle of community life. Chimpanzees are social primates that live in complex groups with frequent and intense social interactions. Like other social primates, they are faced with challenges on two broad fronts. On the one hand, they must attend to the web of relationships with other members of their social group. On the other hand, they must attend to the world outside of their own community by finding food, fending off predators and—most significantly for our inquiry—dealing with members of other communities.

Social animals are often *xenophobic*. That is, they are hostile to individuals living outside the boundaries of their tightly knit groups. Strangers are not just given the cold shoulder. An outsider who strays onto someone else's territory may be viciously attacked and even killed. Contact with a stranger of the same species is the most potent trigger for aggressive behavior among nonhuman animals, and it has been observed in virtually every species with complex social lives.[19]

Social mammals vie with one another for status and power, grovel before their superiors, and attack outsiders without mercy. There is an obvious resemblance to our own long history of social stratification, bigotry, and xenophobia. True to our primate heritage we are hierarchical, ethnocentric animals. Turning, now, to the behavior of our chimpanzee cousins, the connection will become even more unnerving.

WARRING PRIMATES

In chapter 1, I quoted Mark Twain's remark that we are the only animals that wage war. Twain was a fierce advocate of Darwin's theory of evolution, but he died before the systematic study of animal behavior—in particular, primate behavior—had gotten under way. We now know that Twain's statement, although substantially correct, needs to be qualified. We are the only creatures that wage war in the narrow, dictionary meaning of the word, but we are not the only ones to wage war in the expanded sense set out in chapter 3. Chimpanzees share this dubious honor with us.

Chimpanzees are intensely status conscious, and live in communities bounded by high-powered alpha males at the top of the social ladder; low-prestige nobodies at the bottom, and everyone else occupying some intermediate rung. Dominant individuals get more food, more space, and more sex with more desirable partners than lower-ranking individuals. They are also, like big shots in the world of human politics, subject to the greatest amount of stress.[20] Since dominance is usually established through victory in aggressive displays, dominant chimpanzees are often the brawniest and most intimidating members of their community. However, some dominants are hereditary aristocrats, whose status come from being the son or daughter of a high-prestige individual, and others get there by wheeling and dealing. These are the upwardly mobile schemers, who use their wits and social skills to get ahead.

Biologists have observed two broad patterns of primate intergroup violence: the baboon pattern and the chimpanzee pattern. The baboon pattern is a group ritualized display, a *show* of ferocity with few if any casualties. It is nicely illustrated by an account of a battle between two communities of olive baboon (*Papio anubis*) at Gilgil, Kenya, observed by the Dutch primatologist J. van der Hoof. The two groups, consisting of 100 to 150 individuals each, faced off on either side of a line about a hundred yards long, making threatening gestures and shrieking at one another. "Practically immediately," recounts van der Hoof, "the whole front surged forward and chased the other party about a 100 metres to the rear. This was accompanied with a massive swell of shrieking barks, in which all the members of the pursuing group took part with full voice." As the assault lost impetus, the lines reformed and soon the other group launched a counterattack, causing their opposite numbers to retreat. This back-and-forth pattern was repeated for over an hour, and ended as the baboons on both sides of the conflict lost interest in the battle and began to disperse. No injuries were incurred. This general pattern is not confined to baboons: similar behavior has been observed in a variety of other primate species. Impressive though it is, the baboon pattern is more bark than bite. Injuries are rare.[21] The same cannot be said for the chimpanzee pattern of intergroup hostility. Prior to the early 1970s chimps were viewed through rose-tinted glasses. They were poster children for the idea of the essentially peaceable character of humanity in the state of nature. However, in 1974 a man named Mzee Hilali Matama, senior field assistant at Jane Goodall's chimpanzee research center in Gombe, Tanzania, witnessed the first of a number of incidents that caused primatologists to revise their views. A group of eight males that had penetrated into their neighbors' territory found Godi, a young male, sitting alone in a tree. Although he tried to flee, Godi was no match for his attackers, who soon caught up with him, held him down and tore at his body, eventually leaving him bleeding and in great pain. Within a few days, Godi was dead. This event marked the beginning of a conflict between two chimpanzee communities, referred to as the Kahama and Kasakela groups. During the four years that passed after the murder of Godi, the Kasakela group slowly and systematically drove the Kahama chimpanzees to extinction in a bloody war of attrition. This was all the more disturbing because these groups were originally part of a larger community that had divided in two. The killers and their victims had formerly been friends and allies. Further observations revealed the modus operandi of chimpanzee warfare. Small bands of males regularly patrol the borders of their range and invade the areas controlled by neighboring groups. In these circumstances chimpanzees behave very differently from the way that they do in other, more relaxed, situations. Normally, chimpanzees are raucous, scattered, and disorganized, but members of primate commando units move cautiously and silently. They proceed in single file and stick closely together, periodically pausing to scan their environment, sometimes climbing to the top of tall trees to survey the surrounding territory. Like seasoned trackers, they carefully study their environment for telltale signs of other chimpanzees. They sniff the ground and examine traces of feces and urine, partially eaten food, and the remains of night nests.[22] When these bands encounter members of neighboring communities who are on their own and therefore defenseless, they often attack and try to kill them. The only strangers that are not treated aggressively by patrolling males are, predictably, young females in estrus. Attacks can be extremely brutal. Here is how primatologist Martin Muller describes an attack by ten chimpanzees on a lone male from a neighboring community.

> The front of the chimpanzee was covered with 30 or 40 puncture wounds and lacerations, the ribs were sticking up out of the rib cage because they had beaten on his chest so hard. They had ripped his trachea out, they had removed his testicles, they had torn off toenails and fingernails. It was clear that some of the males had held him down, while the others attacked.[23]

Primatologist Richard Wrangham and science-writer Dale Peterson inform us in their book *Demonic Males* that:

> Based on chimpanzees' alert, enthusiastic behavior, these raids are exciting events for them. And the mayhem visited on their victims looks a world apart from the occasional violence that erupts during a squabble between members of the same community. During these raids on other

> communities, the attackers do as they do while hunting . . . except that the target "prey" is a member of their own species. And their assaults . . . are marked by a gratuitous cruelty—tearing off pieces of skin, for example, twisting limbs until they break, or drinking a victim's blood—reminiscent of acts that among humans are unspeakable crimes during peacetime and atrocities during war.[24]

These attacks are very different from the elaborate apparatus of modern warfare, but they are similar to the forms of warfare that are characteristic of hunter-gatherer societies and, by extrapolation, our Stone Age ancestors. What is most remarkable about chimpanzee behavior is the extreme aggressiveness of the attacks and the *intent to kill* that appears to motivate them.[25]

The similarities between chimpanzee and human behavior may cast light on our origins. Human beings and chimpanzees are descended from a common ancestor who knuckle-walked the earth around five million years ago. We don't know a lot about this ancient ape's behavior, but we can make some inferences by attending to the similarities among its descendants.[26] There are many striking parallels between the social behavior of humans and chimpanzees. We both live in complex, highly stratified societies with high-status leaders at the top. Human leaders, like their chimpanzee counterparts, have privileged access to resources, and the alpha members of both species acquire their status either by their sheer impressiveness, by inheriting it from their high-ranking parents, or through political wheeling and dealing. Like chimpanzees, we are extremely xenophobic, and are prone to hostility toward strangers. This suggests that these traits are imbedded in the genetic core of both human and chimpanzee nature.

Comparisons between chimpanzee and human behavior are complicated by the fact that there are actually two species of chimpanzee, the common chimpanzee (*Pan troglodytes*) and the bonobo (*Pan paniscus*) that are, in some respects, as different as night and day. The slender, attractive bonobos are best known for their enthusiasm for recreational sex (including tongue kissing, oral sex, and homosexual play). They form strong female-to-female bonds, whereas chimpanzee females do not form strong bonds with one another. Powerful males dominate chimpanzee females, while males and females are equally dominant among bonobos. Chimpanzee males sometimes physically abuse females, force them to copulate, and kill their infants, whereas no such nastiness has ever been reported of bonobo males. Bonobos live mostly on fruit and only occasionally eat small animals, whereas common chimpanzees regularly kill and eat small game such as monkeys, wild pigs, and antelopes, which they hunt in groups. Most significantly, unlike common chimpanzees, bonobos are tolerant toward members of other communities.

Chimpanzees and bonobos diverged from a common ancestor about two and a half million years ago. A lot hangs on whether the trunk from which the two branches grew was chimpanzee-like or bonobo-like. The ancestor shared by bonobos and chimpanzees had an ancestor at the root of our own lineage. So, if the prehistoric ape that gave rise to the human and chimpanzee-bonobo lines was more like the sensual, affable bonobo than the violent, patriarchal chimpanzee, this might indicate that the heart of human nature is more gentle than truculent.[27] The weight of evidence does not support this uplifting notion of the human pedigree. First, it is inconsistent with a great deal of what we can easily observe about our own behavior. A look at the hierarchical organization of human societies and the extreme violence and xenophobia of human groups shows that, unfortunate though it is, we resemble chimpanzees far more closely than we do bonobos. Second, although we are genetically equally related to both species, there is a wealth of scientific evidence supporting the chimpanzee model of human ancestry. Since the chimpanzees and gorillas had a common ancestor, we would expect the progenitor of the chimpanzee line (which, you will recall, was also the founding member of the human lineage) to resemble a gorilla. Chimpanzees are more like gorillas than bonobos are. Their physical appearance, blood groups, calls, growth patterns, and specific anatomical details are much more gorilla-like than their bonobo counterparts. This suggests that bonobos are something of a novelty, a side road off the main highway of chimpanzee evolution. "Bonobos," writes primatologist Richard Wrangham, "are best imagined as descended from a chimpanzee-like ancestor, rather than chimpanzees from a bonobo-like ancestor."[28]

WHY WAR?

The study of chimpanzee behavior can help us answer the question with which Einstein and Freud wrestled in *Why War?* This ostensibly simple question can be understood in at least three logically distinct but interrelated ways. We can read it as question about the factors that precipitate war—its triggers. What is it "out there" that trips the switch leading to mass violence? We can also interpret it as a strictly biological question. How, if at all, did war contribute to our ancestors' reproductive success? We can interpret it as a psychological question. What is it about the human psyche that makes war possible? In effect, to ask "Why war?" is to ask three questions which have three answers. The answer to the first question is obvious. Chimpanzees attack one another to secure and protect resources for themselves and their kin. By intimidating or exterminating their rivals, a group of chimpanzees can get access to territory and absorb fertile females into the victorious group. Raiding also preempts raids by rival community keen on territorial expansion. The same concerns motivate human war, both on the level of small-scale tribal raiding and large-scale battle. Aggressors seek resources—be they territory, fossil fuels, or religious or ideological converts. In our case, the picture is complicated by several factors, not the least of which is how often fantasy confounds our perceptions of reality. Chimpanzees do not fight over merely imagined fruit trees. How different they are from the Spanish conquistadores, who eagerly hacked their way through the high civilizations of Mesoamerica in search of the mythical El Dorado, the City of Gold. Human beings fight both for resources and for *ideas* of resources.[29] Even intangible goods like "national honor" can in the last analysis be understood as shoring up the resource-holding power of the aggressor.

The second question is also easy to answer. Mark Twain concisely described the principle almost a century ago: "We're nothing but a ragbag of disappeared ancestors."[30] We inherited our warlike nature from prehistoric bands that were able to kill their neighbors and acquire their resources. These groups flourished while the pacifists withered on the evolutionary vine. Another likely reason why war became part of our nature is its intimate relationship with sex. Not only do warriors acquire females as booty, warrior heroes also have an aura of glamour that makes them especially desirable mates. As a result, they have mating opportunities denied to other men. A penchant for war enhanced men's reproductive success, which is why it was selected into our behavioral repertoire.

A common but rather naïve objection to this hypothesis is that war could not possibly enhance reproductive success because it kills young men, and dead men can't reproduce. This reasoning displays an almost total misunderstanding of how natural selection works. It's the winners, not the losers, who benefit from war. To paraphrase Winston Churchill, "Genetic history is written by the victors."

Now, what about the third question—the one about the psychology of war? This one is more complex than the previous two. In a sense, this entire book is an attempt to answer it. For starters, it is essential to distinguish the adaptive function of war from its motives. Although the function of war may be to win resources, that is not necessarily what soldiers have in mind when they march off to battle. As Wrangham insightfully remarks, the tendency to attack one's neighbors "can, in principle, be maintained merely by the tendency of coalitions of successfully raiding males to benefit reproductively, even in unpredictable ways, by weakening the group of neighbors."[31] What matters biologically are the effects of war on individuals' success in spreading their genes. If war enhanced our ancestors' reproductive success (as it almost certainly did) that's enough to account for its continued existence today. This is very different from saying that men go to war because they *want* to spread their genes, because the biological function of a form of behavior is not the same as its motive. Analogously, the function of the twiglike appearance of a walking stick's body is to protect it from predatory birds. But that it not to say that this is its motive. A walking stick doesn't have any motive for looking like a twig. In fact, it doesn't even *know* that it looks like a twig. Our case is slightly different, because there is a relationship—albeit an indirect one—between the biological function of war and its motives. This is pointed out by Wrangham, who goes on to say that "this hypothesis proposes that selection has favored, in chimpanzees and humans, a brain that, in appropriate circumstances, seeks out opportunities to impose violence on neighbors."[32] These considerations give us a very rudimentary and vague explanation of the place of war in human nature (as is indicated by Wrangham's phrase "in appropriate circumstances").

We need to get a clearer idea of the psychological and biological features of human nature that make it possible for people to go to war. Once we have accomplished this, it may be possible to think intelligently about how we can constrain these warlike tendencies that have blighted human life for so long and which may, if left unchecked, drive us to self-extinction.

DEMONIC FEMALES?

"Scarcely a human being in the course of history," observed Virginia Woolf, "has fallen to a woman's rifle."[33] Woolf was right, and her comment suggests that there may be something disingenuous about explaining war as an expression of *human* nature. Might it not be more accurate to say that war is an expression of human *male* nature? Doing battle is, and as far as we know has always been, an almost exclusively male activity.

Women have never been entirely absent from the combat, and there are examples of female warriors to be found throughout history. Queen Boadicea led a coalition of Celtic tribes against the Roman occupiers of Britain. She put London to the torch and slaughtered its 80,000-strong population. The seventeen-year-old visionary Jeanne D'Arc led French forces in a series of stunning victories against

Burgundian troops during the Hundred Years War. These female military leaders often presented themselves as men. This was not because they were particularly masculine. Jeanne had short hair and wore male military dress to avoid the danger of rape. Women like Lucy Brewer who fought in the War of 1812 and Sarah Rosetta Wakeman who fought in the American Civil War, cropped their hair and exchanged their skirts for breeches to conceal their gender from intolerant male eyes. All-female combat units were formed during the bloody Taipieng rebellion, and Russian women took to the field in World War I, although they were a tiny minority of the total combat forces. A larger number of Soviet women were mobilized during World War II, but when female recruitment was at its peak, only about 8 percent of the Soviet armed forces were women. Although most of these were medical and support workers (around 40 percent Soviet military doctors were women), doctors on the front lines also engaged in combat. The Soviets had three all-female bomber regiments and established a women's school for sniper training. Notwithstanding this, it is estimated that in all of history, less than 1 percent of warriors have been female.[34]

Perhaps the universal gendering of war is nothing but an expression of male privilege, and women have been underrepresented in the armed forces because they have been excluded by men. This is certainly part of the truth. Women have faced discrimination in this sphere just as they have in many other arenas. However, to understand it, we must look deeper. Even in the more egalitarian societies like the Vanatinai Islanders, the Semai, Siriono, the Mbuti, the !Kung, and the Copper Eskimo armed warfare is restricted to males.

In all of history, there is only one uncontroversial example of a large fighting force consisting of women. The Gbeto were an all-female elite fighting unit of the West African kingdom of Dahomey (present-day Benin) during the eighteenth and nineteenth centuries. Although never a majority of the kingdom's forces, the Gbeto may have constituted as much as 10 percent of the total number of soldiers in this highly militaristic state. These formidable warriors had a reputation for toughness and ferocity. They were forbidden to have sex with men, were completely segregated from male soldiers, and, according to one European account, were serviced by a legion of female prostitutes. Historians have long puzzled over how to explain this departure from the military norm. The most convincing solution is that the kingdom of Dahomey included women in its combat forces because of falling numbers of men. Dahomey was the dominant power in West Africa, and its economy depended almost entirely on the slave trade. The kingdom was perpetually at war with its neighbors to capture slaves from them, who were then traded to European merchants for rifles, which Dahomeans used to attack their neighbors and capture more slaves in an escalating spiral of war and oppression. Many men died in this incessant warfare, and the king also sold his own subjects to Portuguese slave merchants. The consequent shortage of manpower may have opened the door to the creation of a female fighting force. Whatever the explanation, the Dahomean example demonstrates that women can participate in warfare every bit as effectively as men.

Although soldiering has always been a masculine enterprise, it is not quite right to think of it as a *privilege*. For much of history, ordinary soldiers were treated only marginally better than slaves were. As far back as 1500 B.C. an ancient Egyptian infantryman complained of the life of the ordinary "grunt" follows:

> Just think of how the soldier is treated. While still a child he is shut up in the barracks. During his training he is always being knocked about. If he makes the least mistake he is beaten, a burning blow on his body, another on his eye, perhaps his head is laid open with a wound, he is battered and bruised with flogging.[35]

As late as Victorian times things had not changed very much. Soldiers were poorly paid, poorly fed, and poorly housed. They had higher rates of tuberculosis and other diseases than the general population, and were terrorized by their superior officers, who could abuse and even kill them with impunity. Drunkenness, insubordination, lateness, having dirty weapons, using foul language, slovenly marching, or even a missing button were punishable by burning, beating, flogging on the soles of the feet ("booting"), or having boiling oil poured on one's back. Flogging with a cat-o'-nine-tails was the most common form of punishment. Men endured hundreds of lashes at a time. The knotted whip removed chunks of flesh from the victim's body, and quickly became coated with coagulated blood, which made each stroke even more painful. Sometimes it was soaked in brine before being administered to increase the agony, and men sometimes died as a consequence of flogging. To complete the gruesome spectacle, the *corpse* was sometimes flogged to a bloody pulp. The humane treatment of soldiers was a recent innovation, and is not universally observed even today.[36]

The final reason why we should open our minds to the idea that men are particularly disposed to combat comes from primatology. Among chimpanzees it is the males, not the females, that raid their neighbors' territory. The similarity between human beings and chimpanzees in this respect suggests that there is an innate, biological factor that prompts males to engage in violent combat that is absent or muted in most females. Perhaps this is just an instance of the general principle that males are *generally* more prone to behave far more violently than females.[37] Just look at the statistics on violent crime. On average, men commit 87 percent of the murders in the United States. In most European countries the percentage is even higher. Men commit over 90 percent of the assaults, and studies consistently show that males are more prone than females to physical violence of all kinds.[38] So, are women essentially more peace-loving than men? This is a popular view that is shared by many feminists and antifeminists alike. Like all of the views that we have surveyed so far, it contains a kernel—but only a kernel—of truth. Four decades of polling the attitudes of men and women toward war shows that men express pro-war attitudes more often than women do; however—and this is quite significant—the differences are nowhere near as large as the difference between men and women with respect to *acts* of violence. Here are a few representative examples. In August 1939, 19 percent of American men and 12 percent of American women believed that the

United States should go to war with Germany. In March 1968, 50 percent of American men and 32 percent of American women believed that the U.S. should escalate their involvement in the war in Vietnam, and in March 1975, 55 percent of American men and 38 percent of women agreed with the proposition that wars are necessary to settle differences between nations. These percentages have neither changed significantly over time, nor do they vary with pro- or antifeminist attitudes, age, or educational level.[39] These statistics indicate that women do not suffer from a shortage of violent sentiments—their endorsement of war is less, but not dramatically less, than that of men. How does the image of the peaceful woman sit with the American mother who, in 1943, petitioned the U.S. government to allow her son to send her an ear cut from a Japanese corpse so that she could nail it to her front door? Or how about the smiling young woman who posed for a glossy photo (ironically, in *Life* magazine) with a Japanese skull that her fiancée had sent her from the Pacific theater?[40] Of course, these two dramatic examples do not provide evidence about the inclinations of the majority, but they do graphically demonstrate bellicosity is not confined to those who get blood on their hands. Drawing on writings produced by women during wartime, University of London historian Joanna Bourke concludes that "women satisfied their aggressive urges by pestering their menfolk to act on their behalf and decimate the enemy."[41]

SEXUAL SELECTION AGAIN

If war is a product of biological evolution, and evolution is driven by reproduction, then men's penchant for bloodletting must have something to do with their sexual relations with women. The key to understanding this connection is captured in a comment by the pacifist and feminist Helen Mana Lucy Swanwick, who "ruefully admitted that, although men made war, they could not have done so had women not been so adoring of their efforts."[42] Transposed to a biological context, Swanwick's remark suggests that the masculine warrior mentality is a sexually selected trait, bred into ancestral men by women who preferred warrior mates. Like an inversion of the plot of Aristophanes' play *Lysistrata*, in which the women of Athens and Sparta take peacemaking into their own hands by depriving their men of sex until they agree to end the Peloponnesian war, Stone Age women may have been especially attracted to warriors. If warriors were preferred mates, this may have caused the genes for warlike behavior to proliferate. This general idea is beautifully, if unwittingly, illustrated by a cartoon that appeared on the front page of the *Women's Journal* during World War I showing a suffragette holding a baby standing next to a fully armed soldier. The soldier says, "Women can't bear arms" to which the suffragette replies, "No! Women bear armies."

War is outwardly driven by the need to acquire and retain valuable resources. Resources can make the difference between well-being and starvation, and groups that are able to monopolize them will be more successful than those that cannot. It is thanks to warriors that groups can seize and retain access to such resources. They are also all that stands between a group and its marauding neighbors, who would cheerfully wipe them off the face of the earth. This is why many societies reward their "heroes" with high social status and grant them privileges denied to ordinary members of society. These include sexual privileges. In most species it is in a male's interest to mate with as many females as possible. In our species, the more women that a man can impregnate, the more reproductively successful he is likely to be. However, the same rule does not apply the other way around. Men are not a limiting resource for women because, no matter how many men a woman has sex with, she can only have one baby at a time. According to the harsh logic of Darwinian selection, this implies that it is advantageous for every man to sexually monopolize as many women as possible. Of course, this sets the stage for intense competition among males.

Mandatory monogamy is not the human norm. It is a recent and highly culture-specific practice. An anthropological survey of over a thousand cultures reveals that only 15 percent of them are strictly monogamous.[43] Even in ostensibly monogamous cultures, the rate of marital infidelity can be so high that it is probably best to consider them monogamous in name only. In all societies that allow polygamy, having more than one wife is a marker of status and wealth, because only a man with abundant resources can afford to support multiple mates. Naturally, then, the most powerful men are likely to have the greatest number of wives. Muhammad is said to have had 16 wives and 6 concubines, a very modest number compared to the biblical King Solomon, who supposedly had 700 wives and 300 concubines. Moulay Ishmael the Bloody, the last Sharifian emperor of Morocco, lived in domestic bliss with 1,000 wives who bore him 888 children.[44] In situations like these, the gap between the sexual haves and the sexual have-nots becomes a gaping canyon.

Now, let's consider all of this from a female perspective. Unlike men, women pay an extortionately high biological price for the act of reproduction. All that men contribute is a few minutes of their time and a couple of squirts of semen, whereas women must put up with increasing disability, nausea, back pain, fatigue, and the risk of diseases like preeclampsia and diabetes. This is followed by an extremely painful and possibly injurious delivery and, in the era before modern medicine, a good chance of dying during childbirth. Even today, about a third of the 4 million American women who give birth every year suffer from some form of medical complication. The high cost of reproduction does not end with birth. The nursing infant drains a woman of energy and precious nutrients (among hunter-gatherers, lactation typically continues for years), and she must also care for a helpless infant until it is old enough to become independent. Becoming pregnant is all too easy and is potentially quite self-destructive. So, in light of the sacrifices involved, it pays for a woman to be highly selective. It is in her best interest to choose a mate who is able to provide resources to offset the costs that she incurs, and who is likely to sire vigorous, healthy, successful offspring. As Darwin pointed out long ago, females normally call the shots when it comes to sex. Although males do their best to impress them, and to intimidate rival suitors, it is usually up to females to do the choosing (rape is a strategy used by undesirable males to override the hegemony of female choice). Given the high status of the warrior, and his resource-getting and resource-holding power, it is hardly surprising that military heroes have often been regarded as a prize catch. Warriors are sexy.

In one seventh-century Arabian poem, the female poet seductively entreats warriors to "Advance and our embraces and softest rugs await you." Zulu women used to run naked in front of departing soldiers, and Bedouin women exposed their breasts while urging their men to fight.[45] In the eighteenth century "the concupiscent young officer became an almost archetypal swain in both literature and reality."

> In the case of the hussars, a form of light cavalry descended from Hungarian irregulars, this was not even limited to officers. Resplendent in fur caps, tight britches, bright tunics with horizontal frogging across the chest, even furtrimmed jackets slung over the shoulder, these swaggering examples of genetic advertising earned a reputation as lovers far more devastating than any performance they gave on the battlefield.[46]

The aura of sexuality surrounding the fighter is not restricted to faraway times and places. During World War II British women threw sexual restraint to the wind in their relationships with the American GIs stationed overseas ("We were not really immoral," quipped one British housewife, "there was a war on."). Back in the States "victory girls" or "cuddle bunnies" prowled for sex with servicemen and described their promiscuity as fulfilling a patriotic duty. Speaking of his experiences in war-torn Yugoslavia, Chris Hedges remarks, "There is a kind of breathless abandon . . . and those who in peacetime would lead conservative and sheltered lives give themselves over to wanton and carnal relationships . . . as men endowed with little more than the power to kill are lionized and desired. . . . The killers and warlords become the object of sexual fantasy."[47] Soldiers also become hypersexual. "Most soldiers," remarked Lieutenant Colonel John Baynes, describing World War I infantrymen, "were ready to have sexual intercourse with almost any woman whenever they could."[48] Consequently, war guarantees good business for prostitutes and high levels of sexually transmitted diseases to servicemen (during World War I almost a quarter of the diseases for which British soldiers were hospitalized were sexually transmitted ones).[49]

RAPE

There is, of course, a far less benign path to reproductive success in warfare: the act of rape. One of the perennial attractions of war is the opportunity to abduct or forcibly copulate with women. From an evolutionary perspective, forcing sex on many women may have provided a biological incentive for warfare by enhancing warriors' reproductive success. Warriors have a twofold sexual advantage. Not only are they especially attractive to women in their own communities, but they can also sexually coerce the wives and daughters of defeated enemies. These two factors may have worked together to provide particularly violent ancestral men with more reproductive opportunities than their more gentle peers. Following the well-worn path of sexual selection, the sons of these unions would have been likely to have inherited the warlike temperament of their fathers, as—generation after generation—the genes for male violence proliferated through the population.

Although every nation prefers to imagine its soldiers as clean-cut, heroic, and resolutely virtuous young men, rape is very common, in fact, one might almost say normal, in warfare. We find references to it in the writings of antiquity. In the Book of Numbers, for example, Moses instructs his troops to deal with defeated Midianites as follows: "Kill every male among the little ones, and kill every woman who has known a man by sleeping with him. But all the young girls who have not known a man by sleeping with him, keep alive for yourselves."[50] Centuries later, the Roman historian Tacitus, describing the destruction of the city of Cremona in 69 B.C., grimly wrote: "Forty thousand armed men forced their way into the city. Neither rank nor years saved the victims from an indiscriminate orgy in which rape alternated with murder and murder with rape."[51] Still later, in the twelfth century A.D.,

Genghis Khan proclaimed that "happiness lies in conquering your enemies, in driving them in front of you, in taking their property, in savoring their despair, in raping their wives and daughters."[52] Examples could be multiplied indefinitely.

Rape occurs in virtually all armed conflicts. In war, the female body becomes a resource to be seized and exploited—a piece of booty. Sometimes it is a matter of isolated incidents, and is contrary to official protocol, but at other times it is overwhelmingly frequent.[53] When the Red Army advanced into Germany in the closing days of the World War II, they brought an orgy of sexual violence with them. One Soviet war correspondent later recalled that "the Russian soldiers were raping every German female from eight to eighty. . . . It was an army of rapists." Whole villages of women preferred suicide to falling into the hands of rampaging Soviet troops. The French-Moroccan troops who fought in Italy in World War II were notorious for raping Italian civilians, who begged the Americans to stop them.[54] Rape can take the form of sexual enslavement. Japan imprisoned up to 200,000 (mainly Korean) "comfort women" in brothels for the pleasure of the Imperial forces. Sexual coercion is also used as a form of torture, to extract information.[55] Sometimes it is official policy. Somewhere between twenty and seventy thousand Bosnian women were raped during the war in Kosovo. Many became pregnant and were imprisoned to keep them from seeking an abortion. At least five thousand Kuwaiti women were raped in the short-lived Iraqi occupation of 1990. Rape was also encouraged in the Rwandan genocide:

> One day an official declared, "A woman on her back has no ethnic group." After those words, men would capture girls and take them to their fields for sex. Many others feared their wives' reproaches and raped the girls right in the middle of the killing in the marshes, without even hiding from their comrades behind the papyrus.[56]

Women are sometimes killed or mutilated after being raped. For example, in the Chinese city of Nanking, where over 300,000 civilians were exterminated by Japanese troops, at least 20,000 women were raped. In a typical scenario a woman would be seized, gang-raped, and then killed. If the practice of killing the victim may seem puzzling from a biological perspective, this is because it is an artifact of modern war. In modern warfare, soldiers are often anxious to conceal the evidence of atrocities. As one soldier put it, "We always stabbed and killed them. Because dead bodies don't talk."[57]

What is true of large-scale military conflicts is equally true of primitive warfare. Tribal warriors often regard women as spoils of war, and raids are sometimes undertaken specifically to capture them. "The social position of captive women varied widely among cultures," writes Harvard University anthropologist Lawrence H. Keeley, "from abject slaves to concubines to secondary wives to full spouses. . . . In situations where ransom or escape were not possible, the treatment of captive young women amounted to rape, whether actual violence was used against them to enforce cohabitation with their captors or was only implicit in their situation."[58]

In summary, we are the products of a long process of biological evolution. We are social primates, and share the xenophobic and violent tendencies of our chimpanzee cousins. Those ancestral men with a taste for slaughter had sexual opportunities denied their more peace-loving peers. Treading the well-worn path of sexual selection, they transmitted a proclivity for male coalitionary violence down the generations until it saturated most of humanity. However, there is much more to the story than a description of human beings as glorified killer apes. There is a great deal of difference between the chimpanzee mind and the human mind, and a vast gulf separating the small-scale brutalities of chimpanzee "war" from the colossal horrors of human carnage. To understand the puzzle of war and its evolutionary roots we will need a more subtle approach that does justice to the unique features of *Homo sapiens* and that encompasses both our astounding ferocity and our horror of bloodshed.

5

HAMLET'S QUESTION

> Human reason . . . is called upon to consider questions, which it cannot decline, as they are presented by its own nature, but which it cannot answer, as they transcend every faculty of the mind.
>
> —IMMANUEL KANT, *CRITIQUE OF PURE REASON*

TO UNDERSTAND WAR AS A HUMAN PHENOMENON we must consider a range of issues that may seem at first glance to be remote from it. War is not just about armies and nations and weapons technology. It is a human phenomenon, and to understand it we have to understand the wellsprings of human behavior. In this chapter, we will explore the relationship between mind and brain and consider what it is about the brain that makes us capable of thought and imagination. Most important, we will consider the idea that our minds are not unitary entities, but are actually an assembly of subminds, a team of semiautonomous units that pull together to produce the full range of human behavior and experience. This analysis will yield a conceptual tool kit which will, when put to work in later chapters, help us understand what it is about human nature that makes it possible for us to slaughter our fellow human beings in war.

Let's begin with Hamlet. Right after his eulogy to the human spirit quoted near the beginning of chapter 3, Hamlet's next words frame a powerful rhetorical question. "And yet, to me," he asks, "what is this quintessence of dust?"[1] Shakespeare's eponymous hero is a deeply troubled man, and his question expresses his despair, but it also packs philosophical punch by alluding to a puzzle that has perplexed humanity for more than two thousand years—the so-called mind-body problem.

Let's unpack Hamlet's question. Every word is carefully chosen. The word "quintessence" has a pedigree extending back to the earliest Greek philosophers, who held that the world is composed of four "elements" that they named air, earth, fire, and water. Three centuries later, Aristotle added a fifth element that he called "aether." Aristotle believed that the ordinary elements are finite, corrupt, and changing, whereas the aether—from which the heavenly bodies are composed—is sublime, pure, changeless, and infinite. When Aristotle's works were eventually translated into Latin, the fifth element became known as the *quinta essentia,* literally the "fifth essence." Medieval alchemists believed that all material substances contain the *quinta essentia,* which they sought to liberate from its earthly bondage. The term finally entered English in the fifteenth century as "quintessence." By Shakespeare's time, it had acquired the further meaning of an exemplar or paragon—a thing which is the perfect example of its kind. "Dust" is the very opposite of the sublime, spiritual quintessence. It is matter and flesh, as in "Dust thou art, and unto dust thou shalt return" from the Book of Genesis and the solemn "ashes to ashes, dust to dust" from the sixteenth-century Book of Common Prayer.[2] "The quintessence of dust" thus points to the conjunction of spirit and matter, mind and body, the sublime and the earthly, in a single, exquisitely crafted Shakespearean metaphor. Are human beings the quintessence of dust—purely material creatures—or are we pure, quintessential spirits housed in material shells? And if the former, is this cause for Hamlet-like despair?

Historically, disputes about the mind-body problem have revolved around two sharply contrasting perspectives. The theory known as "dualism" holds that every human being possesses an immaterial mind or soul that is only temporarily associated with his or her physical body and which survives the latter's demise. The other, called "materialism," insists that we are physical beings through and through and that our mental states are nothing but the activity of our brains. The mind-body problem raises some of the most fundamental questions that it is possible to ask about what it means to be human, and the answers that we give have profound ramifications for how we understand ourselves. If we are immaterial beings, as the dualists assert, we are not subject to the natural laws that govern the behavior of the physical universe. We stand apart from the natural world, and our essence cannot be captured in the explanatory formulae of science. If, on the other hand, materialism is true, and the human mind can be fully explained in terms of brain function, science must be the royal road to understanding ourselves. Materialism is attractive because of its promise of knowledge and power and its relationship with the developing sciences of the mind. However, it is also humbling if not demeaning. Many instinctively recoil from a philosophy that seems to demote human beings to the level of material objects.

Until very recently, dualism was the only game in town. For thousands of years, it seemed obvious there must be some invisible, spiritual spark that turns a mere body into a sentient human being. "What," asked Plato "is it that, when present in a body, makes it living?" His answer was "a soul."[3] More than two thousand years later the great French philosopher-scientist René Descartes proposed that human beings consist of a nonphysical mind invisibly hitched to a physical body. Descartes held that the pineal gland, a pea-sized structure located deep in the brain, is the conduit between mind and body, passing messages back and forth between the material brain and the spiritual realm of the soul. The undisputed reign of dualism began to crumble during the seventeenth century, and, as the Enlightenment gathered momentum, opinion shifted toward a more materialistic point of view. Thinkers like the English philosopher Thomas Hobbes and French physician Pierre Cabanis suggested that mental states are actually material states of the brain, and by the middle of the nineteenth century, the new disciplines of neuroscience, psychology, and evolutionary biology were pushing dualism to its breaking point. No matter how carefully investigators searched, they could find no evidence of an immaterial mind pulling the strings that manipulate human behavior. Instead, their research revealed more and more complex *physical* circuitry. Dualism was increasingly seen to be full of holes, and by the middle of the twentieth century most serious thinkers had abandoned what was obviously a sinking ship.

Today, scientists of the mind proceed on the basis that our mental life is produced by processes occurring inside a physical organ—the brain. Like all other macroscopic objects, we are a composite of simple units. We are like elaborate buildings made out of many millions of tiny bricks and can be analyzed into our essential ingredients. *This analysis is exhaustive.* No extra ingredient of mind, soul, or spirit is added to the mix to make it into a human being. It's all a matter of the manner in which these purely physical elements are combined and organized.

If the mind is the brain, as all of the evidence suggests, we are left with the daunting question of how a gnarled ball of neurons is able to produce such a fantastic panoply of thought, imagination, and culture. How can a soggy piece of meat, however complexly organized, formulate the theory of relativity, compose *La Bohème*, or savor a glass of Château Montrose? We do not have anything like an adequate answer to this question, and it is possible that we never will. Understanding how brains make minds may lie as far beyond our reach as differential calculus is beyond the capabilities of a goldfish.[4] But the fact that we cannot yet fully explain how mental states arise in the brain does not mean that there has to be something transcendental or mysterious at work. We have some very good partial and approximate answers to many of the fundamental questions, and our understanding of the neurophysiology of the mind is rapidly advancing. As things stand, there is no credible alternative to a materialistic conception of the mind.

RENAISSANCE BRAINS OR IDIOTS SAVANTS?

A typical human brain consists of just under one hundred billion nerve cells, each of which is connected to up to ten thousand others. Our brains buzz with activity as each member of the immense network of neurons sends and receives electrical signals to and from its neighbors, generating constantly changing patterns of activation. This web of connectivity yields a staggering number of permutations. Multiplying the number of neurons by the number of connections, and making a conservative assumption about the repertoire of possible states for any individual neuron, University of California philosopher Paul Churchland calculates the normal human brain has access to approximately $10^{100,000,000,000,000}$ distinct activation patterns.[5] To get a sense of the sheer magnitude of this number, consider that it is many trillions of times greater than the number of atoms in the observable universe, a measly 1.5×10^{79}. The secret of how this tangle of neurons generates mental states lies in how they are arranged, and what these arrangements are for. Consider, as an analogy, Michelangelo's frescos on the ceiling of the Sistine Chapel. From one perspective, they are nothing but thousands of daubs of paint arranged on a bare expanse of plaster. The power of these little daubs of paint lies in how they are arranged and what purposes they serve. Michelangelo lovingly organized hundreds of thousands of minute smears from his palette in a two-dimensional array representing characters and events from the Bible. Likewise, the tiny nerve cells that compose our brains fire in ever-changing dynamic configurations that represent the world around us. A thought is a picture painted in patterns of electrochemical activation vectors on a canvas of nerve tissue.[6]

Normal human beings have a wide range of abilities at their disposal. Any adult of sound mind can count, comprehend language, discriminate colors, flirt, enjoy music, understand a joke, be embarrassed, empathize with the feelings of others, find their way home, and so forth. We usually take this versatility for granted without pausing to consider how incredibly impressive it is. How can it be explained? One option is to suppose that the brain is an immensely flexible all-purpose organ—a sort of polymath or "Renaissance man" that excels in a great many fields simultaneously. However, over the last two decades, many cognitive scientists have embraced a strikingly different, and more biologically realistic, theory that states that our hugely complex brains are composed of hundreds or even thousands of miniature "brains" that they call "mental modules." The mind is massively modular: an assembly of blinkered specialists, each of whom is brilliant at performing tasks that fall within its area of responsibility, but unable to handle anything outside it. It is more like a collection of idiots savants than a single genius with an encyclopedic knowledge of reality.

Modules are systems for making inferences about the world. "In this model," writes cognitive anthropologist Pascal Boyer, ". . . what makes our minds smart is . . . the fact that very specialized systems are selectively turned on and off when we consider different kinds of objects."[7] The module tweaked when you lay eyes on a captivating member of the opposite sex is quite distinct from the one that is activated on when you discover a large, hairy spider in your underwear drawer or get a whiff of a decaying carcass. Once a module is switched on, action-guiding inference systems swing into action. For example, in the first of these hypothetical cases, the mind is automatically galvanized to scan for cues indicating sexual interest and availability, and then, depending on the outcome, "flirting circuits" may be aroused to steer one's body in a direction likely to maximize sexual opportunity. Obviously, a very different sequence of mental processes would be sparked by an intimate encounter with a tarantula or a heap of roadkill.[8]

Our brains *portray* the world around us. How does this happen? How does monochrome nerve tissue churn out Technicolor experience? Perhaps, one might think, the brain operates like a digital camera, taking snapshots of the external world and then manipulating them with the help of some high-tech neurological software. But this can't be right. The brain does not *literally* contain images. When you think of Abraham Lincoln, there is nothing in your brain that actually *looks* like a tall, gaunt man with a beard. When you get a yen for a piña colada there is nothing in your brain that tastes or smells like coconut, rum, and pineapple. No, the brain represents the world by constructing *models* of reality. "We all have a model builder laboring away in our heads," notes philosopher Colin

McGinn, "working with whatever raw materials lie within the confines of the skull; more soberly put, the brain has the capacity to generate structures that imitate external states of affairs. It is these structures that make thinking possible."[9] To integrate McGinn's metaphor with the modular theory of mind, we can picture it as a huge complex of interconnected workshops, each of which is staffed with a team of highly specialized engineers whose job is to build and continually refine models of the domains for which they are responsible. The layout of this sprawling complex of cognitive workshops is innate: we are born with the basic structure, a template that the process of learning later fills out.

The notion of autonomous units, or homunculi (little men), at work in the brain has been ridiculed for centuries, sometimes rightly and sometimes wrongly. The seventeenth-century German philosopher Gottfried Leibniz parodied the notion that there are "little demons or imps which can without ado perform whatever is wanted, as though pocket watches told the time by a certain horological faculty without needing wheels, or as though mills crushed grain by a fractive faculty without needing anything in the way of millstones."[10] Leibniz thought that this kind of explanation is circular. Explaining vision by invoking a "visual faculty" in the brain would be as vacuous as explaining massive unemployment by the fact that lots of people are out of work, because the so-called explanation tacitly appeals to the very phenomenon that it is supposed to explain. However, the engineers at work in the brain are not Leibnizian imps, and modular explanation is not intrinsically circular. Little agents in the brain are acceptable if they satisfy two criteria: They must be less complex than the mental processes that they are intended to explain, and they must be capable of being factored out or "decomposed" into simpler components, ultimately right down to the level of individual neurons. So although it is inane to chalk up vision to a visual module simpliciter, there is nothing wrong with explaining it as the cumulative output of dedicated neural systems that detect edges, color, motion, and so on, and which can analyzed into simpler subsystems, sub-subsystems and, in the end, individual neurons.[11]

POSSIBLE WORLDS

So far, I have discussed the brain as though it passively represents the world, but there is much more to our mental life than this. "A brain," writes philosopher Daniel C. Dennett, "with its banks of sensory inputs and motor outputs, is a localized device for mining the past environment for information that can then be refined into the gold of good expectations about the future."[12] When naval architects design a ship, they use computer simulations to predict what will happen when the real ship is placed in the water. Tinkering with an image on a computer screen is easier, quicker, and less costly than experimenting on a massive steel vessel in a shipyard. Similarly, when we imagine the future consequences of a course of action, we are reducing our risks by running a simulation on the inner screen of consciousness. Imagination allows us to test-drive our options with no commitment to buy. Faced with an inscrutable and dangerous world, it was hugely advantageous for our ancestors to be able make the mental leap from the seen to the unseen, from the known to the unknown. The great David Hume remarked in his *Natural History of Religion:*

> We are placed in this world, as in a great theatre, where the true springs and clauses of every event are entirely concealed from us. . . . We hang in perpetual suspence between life and death, health and sickness, plenty and want, which are distributed amongst the human species by secret and unknown causes, whose operation is oft unexpected, and always unaccountable.[13]

Our brains needed to be able to do more than simply create models of the world. The power of imagination took root in the brains of our ancestors because it helped them predict uncertain futures. "If as seems evident, the main business of nervous systems is to allow the organism to move so as to facilitate feeding, avoid predators, and in general survive long enough to reproduce," writes University of California neurophilosopher Patricia Smith Churchland, "then an important job of cognition is to make *predictions* that guide actions. . . ."

> *When* an offspring happens to have a mutation in its genes that dictates a structural change in the nervous system that gives the organism a perceptual capacity that allows it to make better predictions than its competitors can make, *then* that organism is more likely to survive and pass its genes on to its offspring.[14]

Today, we perform thought experiments from the safety of our armchairs, fashioning imaginary worlds that test the boundaries of possibility and anticipate the consequences likely to follow from our actions. Flights of imagination empower us to grasp reality all the more effectively, as beings able to see beyond the actual to the possible are better equipped to handle the exigencies of life than those who are chained to the merely actual. Imagination allows us to make inductive inferences or "predictions" about the future. Good predictions are not blind leaps of faith; they are well informed by prior experience. Prior experience comes in four packages. A great deal of it comes from laborious trial and error learning in the School of Hard Knocks. Thankfully, we also have the smarts to learn vicariously from the failures and successes of the people around us. The gift of language makes it possible for individuals to pool experiences. It enables us to learn from others without observing them in action. In addition to all of the practical wisdom that we squeeze out of experience, we each possess a legacy inherited from long-departed ancestors. Thanks to them, we have an instinctive fear of large carnivores, an aversion to sources of contamination and disease, an attraction to fecund landscapes and fertile mates, and many other traits that give human life its distinctive stamp. We extrapolate from these sources of information to chart our course through future possible worlds. This evolutionary legacy provides the floor plan of the structure of our minds, the rooms of which are furnished by the products of personal and vicarious learning. We not only draw on our past experiences, and the experiences of those whom we know, to anticipate the future. We draw on the experience of the entire species that is reflected in the throng of cognitive adaptations that we call our mind.

I have emphasized the constructive side of imagination, but it also has a more sinister aspect. Imagination does not just give us a better grasp of reality. It can lead us badly astray, tearing holes in the fabric of reality and causing us to see things that are not really there. Do you remember being afraid of the dark as a child, cowering under the covers to protect yourself from shadows that turned out to be your bedroom furniture? This is a trivial example of the same process that was at work when six million Jews were exterminated because they were believed to be harbingers of evil, and when nineteen men conspired to smash airliners into urban targets, spurred on by fantasies of rewards in heaven. Imagination is an important ingredient in the psychological brew that leads to war, but before I can explain how, I will have to set the stage a little bit more.

LEIBNIZ'S MILL

Mental processes that unfold outside of awareness underpin all human behavior. When we perform simple actions, like catching a ball, we manage them effortlessly. However, the deceptively simple act of catching a ball depends on a sequence of stunningly complicated computations. You have to work out the ball's trajectory, calculate its speed, and then use this information to predict just when and where your hands need to be, and therefore how your arms need to move, to catch it. All of this has to take place in the split second between the moment when the ball is tossed and the moment that your fingers close around it. If I tossed you a ball and then asked you how you caught it, you would be at a loss for words. At best, you might mumble "I stuck my hands out," or something else equally uninformative. You would be unable to tell me how you caught the ball because the mental processes that guided your actions were *unconscious*. You would be in exactly the same boat if I were to ask you to explain how you tie your shoelaces, how you remember your best friend's name, how you put words together into grammatically correct sentences, and how you manage all of the other little accomplishments of daily life. These actions flow from processes that take place in the unconscious depths of the mind. No matter how hard you try, you cannot be aware of them. Thanks to cognitive science, though, we know that when a person decides to catch a ball, the little engineers in their brain snap into action. The modules responsible for tracking moving objects analyze a thin timeslice of perceptual information and use this to draw up a model of the ball's speed and trajectory. They set your arm and hand in motion, and—using a stream of perceptual feedback—steer them into just the right position. Similar processes lie behind all of our actions and experiences.

To explain why our mental modules do their work outside of awareness, I will help myself to an old philosophical thought experiment. In a remarkable flight of creative imagination for someone living over two hundred years before the invention of the digital computer, Leibniz (the same Leibniz who ridiculed the idea of "little demons or imps" in the mind) asked his readers to suppose "there were a machine, so constructed as to think, feel, and have perception" and that "it might be conceived as increased in size, while keeping the same proportions, so that one might go into it as into a mill." Anyone entering the machine, he rightly says, would see nothing but machinery in action, "only pieces working upon one another," but then he concludes that: "Never would he find anything to explain perception. Thus it is in a simple substance, and not in a compound or in a machine, that perception must be sought for."[15] Brilliant though Leibniz was, he recoiled from the implications of his own thinking. He lost his nerve, and made a blatantly illogical move. Instead of concluding that consciousness cannot be the upshot of a purely physical device, he should have drawn the opposite lesson, for if a conscious machine contains nothing but mechanical parts, than its consciousness *must* reside in those parts "working on one another." Leibniz wrote in the seventeenth century, long before the birth of neuroscience. Today we know that the brain is just such a machine. There are still many people who find Leibniz's reasoning attractive, and who deny that the physical architecture of the brain can account for our mental life. These modern dualists deny that conscious experience can be reduced to physical processes in the brain and claim that there is an unbridgeable chasm between the objective realm of brain events—the material realm of neurons, glial cells, action potentials, and neurotransmitters—and the more rarefied domain of human subjectivity.

As intuitively appealing as this might seem, a slight modification to Leibniz's thought experiment shows why it is fallacious. Imagine the same scene that Leibniz described, only this time imagine that it revolves around a washing machine instead of a conscious machine. Expanding the machine to the size of a mill, you stroll around inside it, observing the motor, fan belt, drum, and various pipes, hoses, and valves, *but you cannot find the washingness!* This is where the great philosopher's reasoning went off the rails. "Washingness" is a silly idea: a bizarre, cooked-up entity. Washing is not a *part* of a washing machine; it is something that it *does*. By the same token, it is a mistake to imagine that there is something in the brain corresponding to our notion of consciousness. Consciousness is not a thing *inside* the brain rubbing shoulders with the anterior cingulate gyrus or tucked away discretely behind the amygdala. Consciousness—if one wants to use this slippery term at all—is something that the brain *does*. The fact that the word "consciousness" is a noun half-seduces us into thinking of it a *thing*. The word "consciousness" should have a verbal equivalent: we should be able to say that the brain is "consciousnessing." Perhaps this could liberate us from the confusions about consciousness that have afflicted philosophers for centuries, and which continue to beguile researchers into the human mind today.

Following on from these considerations, it is easy to understand why our mental modules operate outside of awareness: We are unconscious of them because they are, like the parts of Leibniz's machine, components of consciousness itself. They define the horizons of experience, but they are not part of its content. The idea that conscious experience is the upshot of the operation of an assembly of unconscious modules is, in essence, an updated version of the theory of mind that the German philosopher Immanuel Kant presented to the world in 1781 in his *Critique of Pure Reason.* Kant argued that human experience is shaped by the structure of our minds, and that the kind of mind that human beings possess therefore determines the way that we perceive the world. We live, as it were, in a mind-shaped world. For example, Kant held that we simply cannot help perceiving the world in terms of cause and effect. When something happens—say, a loud bang comes from the room next door—we *never* assume that the sound occurred for no reason; we automatically wonder what caused it. Kant explained that the concepts of cause and effect are built into the fabric of human cognition, and that they color all of our perceptions and judgments, pointing out that it would be impossible for us to learn about causation from experience, because all that we ever perceive are events following other events. We never actually *see* the causal cement that supposedly glues them together. We do not learn about cause and effect by observation: our minds are designed to *interpret* sequences of events as causes and effects. Our picture of the world is filtered through the lens of a concept of causation, a concept that is built into the very structure of our minds.[16]

Contemporary cognitive science supports a broadly Kantian concept of the mind. The "shape" of the mind *does* determine the shape of experience. It also determines our tastes and preferences. It determines the kind of emotions that well up in us, and the configuration of our behavior. Writing in the eighteenth century, long before the science of psychology had taken shape, Kant's account was of necessity speculative, but now it is possible to harness the methods of experimental science to the project of determining how the organization of our brains shapes our experience of the world. Whereas Kant had no way to understand how and why the human mind acquired its specific design, today we know that the mind supervenes upon the brain, and that the brain is a product of our ancestors' long trek through evolutionary time.

Brains are as much the products of evolution as any other bodily organ. As such, they are assemblies of adaptations that exist today because of their positive impact on the lives of ancestral populations. Diverse species have diverse cognitive abilities because their ancestors had to build brains to cope with different types of challenges. The ancestors of modern nocturnal bats had to work out a way of locating small flying insects in the dark. They developed brains with a sonar-interpreting module, a neural system specialized to interpret the faint echoes produced when a bat's high-pitched squeal bounces off a moth in flight. The ancestors of modern eels had to muster solutions to a very different adaptive problems, one of which was the task of finding their way through thousands of miles of open water to a spot in the Atlantic Ocean called the Sargasso Sea, where they breed. Ancestral eels thus had to evolve brains with masterful navigational abilities.

Like bats and eels, our prehistoric ancestors had to evolve special cognitive systems to help them deal with the adaptive challenges that they regularly encountered. They had to be able to acquire mates, find food and water, avoid predators and disease, raise their young, and perform numerous other vital tasks, and they needed brains equipped with modules suited to dealing with these problems. Over time, their brains became our brains, and we inherited the remarkable abilities that enabled them to thrive. This legacy is a mixed blessing. It allows us to perform wonderful feats, but it also mires us in difficulties, some of which are crucial for understanding our passion for war. One of these is self-deception, our uniquely human capacity to hide the truth from ourselves. This will be the subject of chapter 6.

6

A LEGACY OF LIES

> Life has not been devised by morality: it wants deception, it lives on deception.
>
> —FRIEDRICH NIETZSCHE, *UNTIMELY MEDITIATIONS*

HUMAN BEINGS ARE ABLE TO LIE to themselves, and this capacity turns out to be crucial for understanding war. In this chapter, we will explore the architecture of self-deception, peek behind the curtain of consciousness to look at the relationship between mental modules, self-deception, and the unconscious mind. Before doing this, though, I need to make at least a prima facie case that self-deception is important for understanding war (I am not yet ready to present the full story, which will emerge in chapter 10).

As with most aspects of the human condition, Mark Twain had something important to say about the relationship between self-deception and war. Twain's ideas are most powerfully expressed in his story "The War Prayer," which describes a nation on the verge of war.[1] The air simmers with excitement: flags wave, fireworks explode, and young recruits march down Main Street in uniform, as friends and family look proudly on. Mass meetings resound with patriotic oratory, and churches echo with sermons preaching devotion to flag and country. On the Sunday before the first troops are due to depart, the residents of an unnamed town gather to hear the farewell sermon. As the pastor delivers his long, fervent prayer, an elderly stranger walks slowly into the church, steps up to the pulpit, and announces that he is a messenger from God. He tells the congregation that they have offered *two* prayers, one spoken aloud and the other whispered silently in their hearts, and that he has been instructed by God to articulate their unspoken prayer.

> O Lord, our Father, our young patriots, idols of our hearts, go forth to battle—be Thou near them! With them—in spirit—we also go forth from the sweet place of our beloved firesides to smite the foe. O Lord, our God, help us to tear their soldiers to bloody shreds with our shells; help us to cover their smiling fields with the pale forms of their patriot dead; help us to drown the thunder of the guns with the shrieks of their wounded, writhing in pain. . . . for our sakes who adore Thee, Lord, blast their hopes, blight their lives, protract their bitter pilgrimage, make heavy their steps, stain the white snow with the blood of their wounded feet! We ask it, in the spirit of love, of Him Who is the Source of Love. . . . Amen.

"It was believed afterward that the man was a lunatic," the story concludes, "because there was no sense in what he said."[2] "The War Prayer" is about the intimate relationship between war and self-deception. It suggests that citizens hide the raw truth about war from themselves, placing emphasis on glory and heroism while ignoring the horror and suffering that war inevitably imposes on innocent human beings. The same topic was addressed by Aldous Huxley in his book *The Olive Tree*, where Huxley observed that "by suppressing and distorting the truth [about war], we protect our sensibilities and preserve our self-esteem. . . ."

> Finding the reality of war too unpleasant to contemplate, we create a verbal alternative to that reality, parallel with it, but in quality quite different from it. That which we contemplate thenceforth is not . . . war as it is in fact, but

> the fiction of war as it exists in our pleasantly falsifying verbiage.[3]

The discourse of war abounds in doublespeak: contorted language deliberately used to insulate our minds from its hellish reality. During the Vietnam War bombing missions were "protective reactions" and massive defoliation was a "resources control program." The war itself was "pacification." (Twain wrote decades earlier in a jeremiad against the American invasion of the Philippines: "We have pacified some thousands of the islanders and buried them, destroyed their fields, burned their villages . . . and . . . subjugated the remaining millions by Benevolent Assimilation, which is the pious new name for the musket.") In Vietnam, dropping bombs was "delivering ordnance," including "soft ordnance," or napalm, and the defoliant chemicals that ravaged the landscape were "weed killers." Today, we do not *kill* people, we "take them out" or "neutralize the target." Killing your own soldiers by mistake is "friendly fire" and dead civilians are "collateral damage" or the "regrettable by-products" of military action. Bombings that blast flesh to bits are "clean, surgical strikes." A rebellion is an "insurgency," Kidnapping a person in order to have them tortured is "extraordinary rendition." Torture is "physical persuasion" or "debriefing." Assassination is "wet work." Deadly land mines are "area denial munitions." Pumping a person full of bullets is "lighting him up." Stalinist killers in the Soviet Union called deaths under torture "nonratified execution supplements." When a sniper kills a man it is "smoke checking a target." Captain Jason Kostal, a former sniper-school commander remarked to *New Yorker* reporter Dan Baum, "We don't talk about 'Engage this person,' 'Engage this guy.' It's always 'Engage that target.'"[4]

Self-deception lubricates the psychological machinery of slaughter, providing balm for an aching conscience. By pulling the wool over our own eyes and colluding with our own deception, we can continue to think of ourselves as compassionate, moral, and pious people while endorsing or participating in the wholesale destruction of other human beings. To comprehend this in any depth, we need to delve more deeply into the dynamics of self-deception and its antithesis, self-knowledge.

KNOW THYSELF?

The quest for self-knowledge has ranked high on the human agenda for at least 2,500 years. Plato wrote in the fourth century B.C. that "know thyself" was inscribed on the entrance to the Temple of Apollo at Delphi, to which pilgrims traveled from all over the Mediterranean world to seek guidance from the famous Oracle. Around six hundred years later, a Roman writer named Diogenes Laertius stated in his popular *Lives of the Eminent Philosophers* that it was Thales of Miletus, a philosopher who flourished in the sixth century before Christ, who coined the Delphic motto.

Diogenes went on to report that Thales believed self-knowledge to be extremely difficult to come by—in fact, that he regarded it as "the most difficult thing in life" to achieve.[5] At first blush, this seems like a very peculiar thing to say. After all, we are far more intimately associated with ourselves than we are with anything else. But in this case common sense leads us badly astray, for, as we have seen, the notion that we have direct access to the workings of our own minds is more myth than reality. The mental processes that enable you to type a sentence, hit a tennis ball, remember a loved one's birthday, or perform virtually any other mental or physical act lie far beyond the reach of self-awareness. They are as alien to your conscious mind as is the workings of your pancreas. In the last chapter, I mounted an argument that some mental processes are unconscious because of the way the brain is configured, but this is not the only reason why it is difficult to know our own minds—not by a long shot. Twain's congregation were unconscious of the implications of their prayer because of a desire to *hide* the truth. Clearly, we need to distinguish form of unconscious mental activity described in Chapter 5 from the kind of unconsciousness illustrated in this story.

"Were a portrait of man to be drawn . . . in which there would be highlighted whatever is most human . . . we should surely place well in the foreground man's enormous capacity for self-deception," wrote the philosopher Herbert Fingarette on the first page of a classic study of self-deception. Many students of human nature, both ancient and modern, have concurred with this assessment.[6] Immanuel Kant described what he called the "hard descent into the Hell of self-knowledge" and claimed that the outcome of self-observation is "the gloomiest melancholia."[7] Kant's suspicions are supported by modern psychological research showing that people who suffer from depression perceive themselves more accurately than so-called mentally healthy people. Kant's younger contemporary, the polymath Johann Wolfgang von Goethe, went even further, condemning "Know thyself" as a "deception . . . to confuse humanity with impossible demands." "Know thyself?" he wrote, "If I knew myself I would run away."[8] And Twain added, in much the same spirit, "Man, 'Know thyself—& then thou wilt despise thyself, to a dead moral certainty.' "[9] The insightfulness and candor of these men provides a refreshing contrast to the unwitting duplicity of those who unthinkingly proclaim that *of course* they want to know themselves. But do we really want to be free of illusions about ourselves? In practice, we seem to want to "know" ourselves only very selectively. We happily embrace a self-portrait with most of the imperfections airbrushed out, but are less happy to accept the unvarnished truth.

A throng of philosophers, playwrights, and novelists have underlined the pervasiveness of our timeless longing for illusion.[10] The Dark Continent of self-deception has even found a place on the research agenda of contemporary psychology. In one study, college students were asked to read an article presenting empirical evidence showing that caffeine consumption causes an increased risk of noncancerous breast lumps in women. When asked how convinced they were that this claim was true, the female coffee swillers were far more skeptical about the relationship between caffeine and fibrocystic disease than those women who consumed caffeinated drinks less frequently. The coffee drinkers did not base their views on an objective, well-reasoned assessment of the evidence in the article. They were skeptical because they didn't *want* caffeine to be a health risk, and indulged in a little wishful thinking to neutralize the threat.[11] More alarmingly, a review of the scientific research into faulty self-assessment concludes that "in general, people's self-views hold only a tenuous to modest relationship with their actual behavior and performance. . . . Indeed, at times, other people's predictions of a person's outcomes prove more accurate than a person's self-predictions."[12] For example, surgical residents' assessments of their skills do not match the objective measures on medical-board exams (the residents tend to overestimate their abilities). College students can accurately predict how long their roommate's romantic relationships will last, but cannot predict the longevity of their own. Lawyers tend to be unrealistically optimistic about their prospects for victory in court, investors make unwarranted assumptions about the moneymaking potential of their investments, and most of us labor under the false belief that we are more adept at assessing ourselves than the person next door.[13]

REPRESSION REDUX

Self-deception seems perplexing. The idea that a single person is both perpetrator and victim of her own deceit can seem so bizarre as to be implausible. For deception to work, the deceiver must be in the know and the deceived must be in the dark, but in cases of self-deception deceiver and deceived are the same person. A deceiver is like a stage magician who knows what makes a trick work, but must prevent his audience from catching on. Self-deception may seem every bit as impossible as a magician concealing the mechanics of a successful trick from *himself*.

If self-deception often appears to be paradoxical, it is because many of us are married to a false conception of the human mind. This outmoded theory was most clearly articulated by René Descartes in the seventeenth century. Descartes reasoned that because the mind is a nonphysical thing, it cannot occupy space. It then occurred to him that only things that occupy space can be split into parts (it makes no sense to say that something that does not have any size can be divided in half). So, if the mind does not occupy space, if it is not a physical thing, it cannot have any parts. Of course, this conclusion only follows if we accept the first premise—the notion that the mind is not physical. If we reject it, and claim instead that the mind is the brain, as I argued in chapter 5, then we do not have to accept that it has no parts. Of *course*, the brain has parts, and some of these are functional parts: the mental modules discussed in the preceding chapter. The mind is constructed from many functional parts: systems, subsystems, and sub-subsystems, right down to the level of individual neurons, and this arrangement gives us the conceptual elbow room we need to divest the idea self-deception of its aura of weirdness. The modular organization of the mind allows for the possibility of *conflict* between its components and, most important, for the possibility that one part of the mind can withhold information from others.[14] It may seem strange to conceive of your brain as an arena in which a number of semiautonomous entities interact with one another.

Although I have used some poetic license to get the point across, the notion that the mind resembles a contingent of inner agents whose interests sometimes converge and sometimes conflict is broadly consistent with recent discoveries in psychology, neuroscience, and even genetics.

If you feel that this is too far-fetched, you might consider some intriguing observations of people who have undergone split-brain neurosurgery. Split-brain surgery is a procedure pioneered during the 1960s to control severe forms of epilepsy. It took advantage of the fact that the brain is naturally divided into two hemispheres, one on the right and one on the left, connected by a highway of densely packed neural cabling known as the corpus callosum. Normally, information flowing back and forth across this bridge permits our brains to function as an integrated whole. However, when one hemisphere is seized by temporal lobe epilepsy, the resulting cascade of uncontrolled neural activity can spread across the corpus callosum like a raging brush fire, and drag the entire brain into an epileptic seizure. Neurosurgeon Roger Sperry, who received a Nobel Prize for his work, realized that it was possible to prevent the spread of epilepsy in the brains of his patients by slicing through the corpus callosum to create a sort of firebreak. The procedure worked, but it produced some bizarre consequences. Sperry and his team found that these individuals sometimes behaved as though they possessed two fully functioning, autonomous minds competing for control of a single body. For example, one patient was observed struggling to pull his pants up with his right hand while at the same time yanking them down with his left. Another assaulted his wife with his left hand while defending her with his right.[15] Splitting the brain appeared to split the mind. Investigating further, Sperry found that that the two sides of the brain are specialized for different tasks. The left hemisphere is normally responsible for analytical and verbal functions, while the right half attends to spatial perception, music, and emotionally charged language like swearing. When our brains are intact these differences are present but are in normal circumstances virtually undetectable. Sperry concluded that, although we are usually unaware of it, all human beings possess a composite brain that contains "two separate realms of conscious awareness; two sensing, perceiving, thinking and remembering systems."[16] According to University of California neuropsychologist Michael Gazzaniga, one of the main functions of the brain's left hemisphere is to make sense of experiences by weaving them into narratives about our lives, narratives that give the appearance of seamless coherence, sometimes at the expense of truth. The left hemisphere's penchant for confabulation was brought sharply into focus by some of Gazzaniga's experiments. In one of these, the word "walk" was presented to a split-brain patient's right hemisphere, and he responded by getting up from his chair and walking across the room. When asked why he did this, the patient replied, "I wanted to get a Coke."[17] His left hemisphere concocted the story about wanting to have a Coke because it was unaware that his right hemisphere had been given the instruction to walk. This is not quite self-deception, but it is awfully close to it. The disjunction between the real motive (the injunction to walk) and the phony explanation of it ("I wanted to get a Coke") was induced by the surgeon's scalpel. It may nevertheless teach us a valuable lesson.

Perhaps *psychological* forces can have effects similar to surgery, setting up informational roadblocks in the brain that prevent it from operating as an integrated whole.

Finally, let us turn to the view from genetics. In the past, when biologists spoke of conflict, they almost always meant conflict for resources played out between whole organisms. But recent research suggests that biological conflict may also unfold *within* organisms, even within the limited confines of their brains. The idea of internal conflict is nothing new. It has long been the province of philosophers and novelists. But now science may be on the verge of revealing precisely what it is about the physical constitution of the human brain that makes these inner dramas possible.

Like shoes, genes come in pairs: each of us inherits one copy from our mother and one from our father, but only one member of each pair can have an impact on the growing organism, one is "expressed" and the other is silenced. Generally speaking, whether or not a gene is expressed has nothing at all to do with whether it comes from your mother or from your father. However, not all genes work in this way. Some are "imprinted"—marked, so to speak, with the stamp of their origin that guarantees that they will be expressed. Maternally imprinted genes are only expressed if they come from your mother, and paternally imprinted genes are expressed only if they come from your father. In practice, this means that your paternally imprinted genes "work" for your father, while your maternally imprinted genes are agents for your mother. This provides fertile ground for the growth of internal conflict, as proxies for mother and father duke it out inside a single person.

One of the best arenas for studying genetic conflict is pregnancy. David Haig, a professor of biology at Harvard University, has described the battles that take place between maternally and paternally imprinted genes in a pregnant woman's body. Paternally imprinted genes are only "interested" in producing a big, strong, healthy baby, even if this drains the mother, so they produce a substance called "insulin-like growth factor 2" (*Igf2*) that makes the embryo grow. Maternally imprinted genes aim to preserve the mother's well-being and produce insulin-like growth factor 2 receptors (*Igf2r*) that mop the stuff up. In experiments with mice that were engineered to develop from genes taken from two males or two females, rather than the normal male-female parentage, the mice with two fathers turn out to be beefy hulks with huge bodies 140 percent of normal birth weight, but they have tiny brains. This happens because in mice with two dads the paternally imprinted growth genes are given free play to pump out *Igf2* but have no maternally produced receptors to put a brake on it. Those with two mothers turn out to be nerdy little rodents with hefty brains but with bodies only 60 percent of normal birth weight: the maternally imprinted genes prevent them from hogging the extra nutrition from their mother.[18]

According to Cambridge behavioral neuroscientist Eric Keverne, the weight of evidence suggests that the cerebral cortex and striatum—parts of the human brain that produce sophisticated cognitive and motor skills—are built from maternally imprinted genes, whereas the more primitive, passionate limbic brain—the part of the brain that drives us to eat, drink, have sex, and attack enemies—comes from their paternally imprinted counterparts. Circumstantial evidence supporting this comes from two genetic disorders. In Angelman syndrome, a baby is born with two copies of a paternal gene at chromosome 15 instead of one from each parent. People suffering from Angelman syndrome are mentally handicapped, make jerky movements, and have speech difficulties, all of which are related to malfunctions in the cerebral cortex and striatum. Prader-Willi syndrome is the opposite: there are two maternal copies of the gene at the same locus on chromosome 15. Sufferers from Prader-Willi syndrome have a poor sucking reflex and a weak cry as infants, and are emotionally tractable and lack sexual drive as adults, all of which suggests defects in the limbic brain.[19] If it is true that separate brain systems are built from maternally and paternally imprinted genes, it is just a short step to the idea that different parts of the brain may have competing interests, and may conflict with and even deceive one another.[20]

A skeptical reader might object at this point that even if all of this is true, the mental divisions suggested by psychologists, neuroscientists, and geneticists are all different. There is no reason to think that the division between the right and left hemispheres uncovered by Sperry and Gazzaniga has any basis in imprinted genes or that they correspond to the modules hypothesized by cognitive scientists. This is a fair criticism, but it misses the point. Even though the results from the three disciplines do not map onto one another, they each provide independent support for the idea that the mind is divided in ways that may play a role in the peculiar phenomenon of self-deception. Nature—including human nature—is not regimented into neatly labeled compartments. It consists of a messy array of overlapping categories, from which we are free to choose the scheme that best suits the particular explanatory task confronting us.

To delve more deeply into how self-deception might work, I now turn, with some trepidation, to Sigmund Freud's theory of the mind. Freudian ideas have come under heavy fire over the last few decades, and psychologists now mostly regard them as unworthy of serious attention. It is true that Freud was an extremely speculative thinker whose conclusions often rested on unstable evidential foundations. He was also a child of the nineteenth century, and many of the cutting-edge theories that he based his thinking on are now little more than quaint anachronisms. Worse, recent scholarship has shown that a number of his clinical claims were, shall we say, less than completely truthful. But these are not good reasons for trashing his work in toto. Freud had a bold and brilliant theoretical imagination, and his writings still repay close consideration. In particular, his conception of the relationship between conscious and unconscious mental states, and the theory of self-deception that flows from it, have weathered rather well: scientific advances have, if anything, rendered it more plausible.[21] Freud believed that human beings conceal socially unacceptable thoughts, feelings, and motives from themselves. This general idea was already deeply entrenched in the Western intellectual tradition by the time Freud entered the fray in the closing decades of the nineteenth century, but he was unique in giving a detailed, coherent account of those features of the mind that make self-deception possible. His theory of self-deception (or "repression," as he preferred to call it) is based on his more general conception of how the mind works and, in particular, the relationship between thinking and consciousness.

Most of us share a broadly similar and largely unarticulated assumption about our conscious mental processes which seems so obviously true and is so completely taken for granted that it is difficult to question. The assumption is that thinking is a conscious process and that we are aware of our thoughts at the very moment that we think them. Freud rejected this assumption. His theoretical reflections suggested to him that all thinking occurs unconsciously and that we become aware of our thoughts only *after* having thought them. If this is difficult to wrap your mind around, consider the operation of a compact disk player. When you want to listen to some music, you select the appropriate polycarbonate disc, place it in the tray of a compact disc player, and press the button marked play. As far as most of us are concerned, that's the end of the story. But in reality, it is just the beginning. Once the play button is pushed, a lot has to happen before you can hear the music. First, a light-emitting diode in the machine must generate a laser beam that is split and then focused by a lens on the disc's surface. Variations in the reflected beam cause the machine to generate a sequence of digital signals. These pass in single file through a digital to analog converter that translates the discrete zeros and ones into a seamless flow of analog information. After amplification, the flow of information makes its way to the music system's speakers, which produce patterns of vibrations in the air. These strike your eardrums causing them to hammer out a sequence of neural signals that are shunted off to the auditory areas of your brain. Eventually, they transduce into the sound of Charles Mingus playing "Stormy Weather" (I have *vastly* simplified the story of the neural processing that has to happen before you can hear the music). Looking closely, an ostensibly simple and immediate process—playing a compact disc —involves a whole series of steps and a certain amount of time. Freud thought that the same is true of the process of becoming conscious of one's own thoughts. Conscious thoughts do not appear out of nowhere; they are produced by unintrospectable events unfolding inside of your brain.

If Freud was right, there must be a brief window in time between the moment that a thought occurs and the moment that the thinker becomes aware of it. In Freud's day this was pure speculation, but today it should be possible to measure the gap using the sophisticated methods that are now available to us. The most striking demonstration of this time lag is a famous experiment conducted by the neuroscientist Benjamin Libet. Libet asked his subjects to perform a simple action, pushing a button, whenever they felt like doing it while he simultaneously monitored the electrical activity in the part of their brain responsible for voluntary movement. Libet discovered that almost a third of a second elapses between the moment that a person decides to perform an action and the moment that they become aware of making the decision![22] If this interpretation of Libet's results is correct, it raises a further question. What goes on in the brain or mind during that mysterious 300 milliseconds? Freud advanced an hypothesis that may throw light on this mystery. I will explore it with the help of a thought experiment. Imagine that you are observing Mary, the chief neuroscientist at Jackson Memorial Hospital, who is about to demonstrate an amazing machine that she has invented. The machine is called a brainoscope, and it can read people's minds by monitoring the electrical activity in their brains. Mary attaches electrodes to the scalp of Frank, her experimental subject, and flips a switch on the 'scope. Lights flash. The brainoscope hums and bleeps as it boots up. Next, a television monitor attached to the machine comes alive with a complex geometrical pattern and a string of numbers moves along the bottom of the screen. Mary turns to you with a knowing look and, gesturing to the screen, says, "See?" But these patterns and numbers are quite meaningless to you. You need her to interpret them. "Oh, I'm terribly sorry," she says, "I forgot that you don't understand the language of neural-activation vectors. This particular pattern means that Frank is thinking about a ripe tomato!" This vignette captures Freud's conception of our relationship with our own mental processes. Freud thought that all mental processes are brain processes, but that these brain processes are not accessible to us as such. We cannot directly observe the neurochemical activities that go on between our ears. But even if we could—say, by placing ourselves in a brainoscope—we would not understand what the brain processes *mean*. If someone were to show you a snapshot of the processes going on in your brain at precisely 11:13:06 EST on February 16, 2005, you would be none the wiser about your mental state at that time. Freud reasoned that in order for us to become aware of our own thoughts, there must be a mechanism that *translates* them into a medium that we can understand. Let's say, for the sake of the argument, that human brains come equipped with a neurological system that transforms neural-activation vectors into information couched in a symbol system that ordinary human beings can make sense of. What form might this translation take? Freud thought that it takes exactly the same form that it took in our thought experiment when Mary explained the readout from her computer. He hypothesized that each of us possesses a module that translates brain activity into language, converting information about signals firing across synapses into a silent monologue that we call conscious thought. This may be what happens during the interval between thought and consciousness: the thought is being translated into the symbolic medium of language so that it can be accessed by the conscious mind.

Freud's thesis is consistent with much of what cognitive science has discovered about the relationship between language and thought.[23] It also gives an elegant model of how self-deception might work. Imagine that the processes described in Freud's theory are unfolding in your brain right now. As you sit, a welter of thoughts are generated deep within your brain which enter your consciousness as a sequence of unspoken sentences—sentences like "I must remember to stop by the supermarket on the way home" or "Freud wasn't so loony after all." Now, consider what would happen if the module that translates neural information into sentences malfunctioned and that the stream of sentences flowing through your consciousness no longer remained faithful to your unconscious thoughts. If this happened, your mental monologue would continue, but it would not express your true thoughts—in fact, *you would no longer be aware of what you are really thinking*. Strangely, you would be none the wiser, for there would be no way for you to compare the thoughts that you believe yourself to be thinking—mental sentences heard with the inner ear—with the unconscious thoughts that they are supposed to express. Freud hypothesized that something like this occurs when we deceive ourselves: the module that translates unconscious thought into conscious self-talk is disrupted by a psychological motive such as fear, guilt, or shame. This disruption creates a state of false consciousness, a cognitive embargo that selectively replaces authentic thoughts with empty proxies, and creates a bastion against self-knowledge.

THE BIOLOGY OF SELF-DECEPTION

Most people regard self-deception as blatantly maladaptive, a symptom of the machinery of the mind having gone awry. This is certainly the standard view in the mental-health profession. But this may be a false assumption. How about the alternative hypothesis that the mind switches into states of self-deception because there are times when it is profitable *not* to know? Ignorance is often bliss, but there are also occasions when remaining in the dark provides a better return than mere enjoyment. If self-deception reaps some advantage for its perpetrator (and ostensible victim) it may be that self-deception is a normal, healthy state rather than a symptom of disorder. Treating it as pathological may itself be a form of self deception, a way of lying to ourselves about the extent to which we lie to ourselves.

Perhaps the very opposite is true, and self-deception is an adaptive, life-affirming feature of the human mind. This presents a puzzle. Selection only favors those traits that benefit an organism's genes, either directly, by increasing its breeding opportunities, or indirectly, by increasing survival prospects. If self-deception is an adaptive trait, we need to understand how and under what circumstances it might be beneficial to deprive oneself of information. It seems obvious common sense that the more information we have at our command, the better we are able to negotiate the hazards and exploit the opportunities presented by an ever-changing environment. But common sense is a poor guide to reality, and progress in scientific understanding typically involves getting under the skin of reality and moving from the merely obvious to the genuinely veridical.

To get at the adaptive role of self-deception, we must first consider deception simpliciter. Why do people deceive one another? An engaging way into this topic is through the writings of Plato, who recounted in the first two books of *The Republic*, how eleven men gathered for an informal get-together one glorious afternoon almost twenty-five hundred years ago at the home of a wealthy merchant in the port of Piraeus, just outside of Athens. The group included the philosopher Socrates, a professional rhetorician named Thrasymachus, and Plato's brothers Glaucon and Adeimantus, among others. After an exchange of pleasantries, the conversation turned to the question of the nature of justice. As the discussion unfolded, Thrasymachus, who was a bit of a bully, argued that just conduct is for simpleminded fools and that the really astute person is ruthlessly self-serving. "You can't avoid the conclusion, my simpleminded Socrates," he sneered, "that a just man comes off worse than an unjust in every situation."

> Take contracts, for a start, where a just man goes into partnership with an unjust. When the partnership is dissolved, you'll never find the just man better off than the unjust. No, he'll be worse off. Or think about public life. When there are special levies to be paid *to* the state, the just man contributes more, and the unjust man less from the same resources. When there are distributions to be made *by* the state, the just man receives nothing while the unjust man makes a fortune. Or suppose each of them holds some public office. The outcome for the just man, even if he suffers no other loss, is that his own financial position deteriorates, since he cannot attend to it, while the fact that he is a just man stops him from getting anything from public funds. On top of this he becomes very unpopular with his friends and acquaintances when he refuses to act unjustly in order to do them a favor. The outcome for the unjust man is the exact opposite.[24]

Thrasymachus's argument contains a gaping hole. Surely, a person who is *overtly* exploitative (say, a politician who obviously feathers his own nest at the citizens' expense) is more likely to be thrown out of office than to garner accolades. The Machiavellian operator needs to hide his machinations behind a smoke screen of deception. This was not lost on Glaucon, who waded into the argument to raise this very point:

> The one who gets caught is to be regarded as incompetent, since perfect injustice consists in appearing to be just when you are not. . . . To the person who commits the greatest wrongs we must not deny—in fact, we must grant—the enjoyment of the greatest reputation for justice. If he makes a false move, we must allow him the ability to put it right. He must be capable of using persuasion—so that if any evidence of his wrongdoings is brought to light, he can talk his way out of it. . . . [25]

Adeimantus then elaborated:

> If I am unjust, but have gained a reputation for justice, then I am promised a wonderful life. Therefore since "Appearance," as the wise men have pointed out to me, "overpowers truth" and controls happiness, I must turn all my attention to that. I must draw an exact likeness of goodness around myself, as a front and a façade. . . . "The trouble with that," someone will say, "is that it is hard to be evil and get away with it for ever." "Well," we shall say, "nothing great was ever easy." But if we are going to be happy, we must follow where the trail of our argument leads us.[26]

Thrasymachus, Glaucon, and Adeimantus tried to analyze social life in terms of intentional strategies. In their accounts, the prototype of the successful man is the person who *deliberately* exploits others to achieve his aims and pulls this off by lying to them. Even today, we often assume that such people must be fully aware of their deceitful maneuvers. We often assume that a good liar must be conscious of his or her dishonesty. But in fact, the very opposite is probably more accurate; the less conscious we are of our own dishonesty, the more effectively we are able to hoodwink others. The best liars are unconscious ones. To understand why this is, we need to have a look at the biology of deception. Cooperative alliances between individuals who trade favors are found throughout nature—both within and between species. Generally speaking, cooperation is a win-win situation. If you hunt in a group, you bring home considerably more bacon than a solitary hunter can secure on his own. If you are defending yourself from a dangerous predator, your survival prospects are hugely improved if you are one member of a united front. If you need to build a shelter, many pairs of hands are better than only two. Imagine a Thrasymachean world where everybody is single-mindedly devoted to his or her own interests and nobody cooperated with anyone else. The denizens of such a world would have very poor prospects. They would be condemned to the sort of grim, impoverished existence portrayed by Thomas Hobbes in *Leviathan:* a world in which people live in "continual fear, and danger of violent death; and the life of man, solitary, poor, nasty, brutish, and short."[27]

If thoroughgoing cooperation and stark selfishness were the only two alternatives, then cooperation would win hands down. But cooperation is not all that it seems, and anyway, as Adeimantus pointed out, appearance trumps reality in our dealings with one another. Sometimes individuals who appear to cooperate are actually cheating, indulging in some sleight of hand and covertly taking more than their fair share. Every year, millions of people do this when completing their income tax forms; they pretend to be declaring their entire income and true deductions while secretly massaging the figures to their own advantage. People cheat, because cheating benefits them. The person who pays less than their fair share of taxes gains an advantage at the expense of those who are more honest. Biologists call this the "free-rider problem." In any cooperative venture, it pays each individual to cheat and to exploit the others for his or her benefit.

However, it is detrimental to *everyone* if everyone cheats, because cooperation falls by the wayside. It follows that it is in each individual's interest to exploit others, while preventing others from exploiting him. You might call this the evolution of the double standard. We preach the paramount importance of honesty and integrity, while giving ourselves considerable wiggle room. Many of the same people who lie on their tax declarations do not hesitate to stand up and lecture others about the virtues of good citizenship.

Cheating is a gamble. It promises profit but threatens serious losses. The obvious way to maximize benefits while keeping losses to a bare minimum is to cheat *secretly*: an efficient cheater is a deceptive cheater, who can violate the trust of others without their realizing it. This allows the cheater to savor the best of both worlds. He maintains an unblemished reputation, and reaps the rewards that flow from it, but he also gets something extra on the side. Those who clandestinely cheat—who fail to reciprocate or who reciprocate only partially—will, all things being equal, do better than those who are scrupulously fair. Nature selects for what *works*. So, if dishonesty works, we should expect to find deception pervading the biosphere. In fact, this is precisely what we find. Nature is an intricate tapestry of guile, and is full of organisms that use every manipulative trick in the book to help them survive and reproduce. Some are masters of camouflage and are able to seamlessly merge with their environments, others imitate inedible or noxious objects to avoid being eaten. Some disguise themselves as members of the opposite sex to avoid costly competition or to finagle expensive courtship gifts, while others pretend to be toxic to discourage predators. Nonhuman deception is all-pervasive and extraordinarily varied.[28]

Homo sapiens is also a deceptive species, and our superb intellect and mastery of language empowers us to deploy tactics infinitely more devious than the techniques available to our nonhuman cousins. Each of us knows that we are vulnerable to being duped by others, and we try to avoid this at all costs. Our background awareness that social life is a complex game of hide-and-seek, of manipulation and countermanipulation, makes us wary. We also know that others are similarly wary of us. This drives up the stakes when we try to lie. We know that we may be found out, and this burden of self-consciousness releases a rising tide of anxiety. Think of a time that you told a serious lie—not a white lie but a lie that *mattered*. Think about how you felt when you were telling it. Your heart was probably pounding in your chest, and your breathing was rapid and shallow. Perhaps your nose began to itch, and you had a sinking feeling in the pit of your stomach. You didn't know what to do with your hands and you began to perspire. The sensation of warmth spreading across your cheeks warned you of an impending blush, and you could not manage to hold the other person's gaze for very long. Worse, your awareness of all of this made you even more anxious and awkward. Now, imagine a married woman arriving home after a session with her lover. Her husband asks her why she is late. She hesitates, blushes, fidgets, averts her gaze and, stumbling over her words, and tells him in a croaky voice that she had been out for a drink with a female friend. How convincing do you think she would be? Which would be more persuasive, the story deliberately told by her mouth or the tale told involuntarily by her body?

This suggests an answer to the riddle of what makes self-deception advantageous. Twain pegged it almost a century ago. "When a person cannot deceive himself," he wrote, "the chances are against his being able to deceive other people."[29] Although we pay lip service to the glories of self-awareness, it may be that there are circumstances in which too much self-awareness is a problem, and self-deception kicks in like a circuit breaker when the voltage of self-awareness becomes dangerously high. This is the essence of a hypothesis proposed by evolutionary biologist Robert L. Trivers in the 1970s. Trivers argued that the capacity for self-deception may have been selected for because it helps us deceive others. A liar who believes his own lies is far more convincing than one who doesn't, so if you sincerely believe that you are telling the truth, there is no reason to be awkward and self-conscious. The accomplished self-deceiver can deliver self-serving falsehoods without even breaking a sweat.[30]

Our capacity for self-deception presents a formidable obstacle to self-knowledge. Consider racism. In this day and age few people are prepared to admit, even to themselves, that they are racially prejudiced. Unfortunately, this may owe more to human duplicity than to the widening circle of brotherly love. Mahzarin Banaji, a social psychologist at Harvard University, uses an experimental technique to peek behind the curtain of consciousness to tap unconscious racial prejudice. Her method is simple. She asks subjects to look at a computer monitor upon which appears a sequence of positively and negatively toned adjectives, each of which is paired with a stereotypically white or black name (for example, "Meghan" and "Keisha"). As each name appears, the person taking the test is asked to press a key to register whether the adjective is good or bad. While this is going on, the computer measures how long it takes the subject to respond to each word. Although one might expect reaction times to each word-name pair to be broadly similar, Banaji found that most people respond most quickly when white names are paired with positive adjectives and black names are paired with negative adjectives. Why is this? Well, the difference in speed suggests that her experimental subjects (and, by extrapolation, most Americans) process the white/good and black/bad pairs more quickly than the white/bad and black/good pairs. Banaji suggests that there is a preexisting link between goodness and whiteness, and badness and blackness in the minds of her subjects, and that their brains quickly assimilate the information to these stereotypes, whereas white/bad and black/good pairs violate the stereotypes and thus require more time to process. Banaji discovered that the same effect in people who claim not to be racially prejudiced, and in African-American as well as Caucasian subjects.[31]

Trivers's theory implies that people deceive themselves about matters disapproved of by their social milieu. Freud, too, thought that self-deception congeals around precisely those areas of human nature that are most vigorously condemned by one's community. Long ago, Freud suggested that men with strong feelings of repugnance and hostility toward homosexuals deceive themselves about their own homoerotic urges. This hypothesis was strikingly confirmed by University of Georgia psychologist Henry E. Adams and his coworkers in what has become a classic in the scientific literature on self-deception. Adams's team asked two groups of men, a homophobic and a nonhomophobic group, to watch films depicting heterosexual, lesbian, and homosexual lovemaking while attached to a plethysmograph, a device that measures subtle fluctuations in blood flow to the penis, and therefore subtle indications of sexual arousal. Although both groups reported being sexually excited by the lesbian and heterosexual films, and these subjective reports were confirmed by the objective measurements, *only the homophobic men* showed signs of being aroused by the homosexual scenes, in spite of their vehement denials.[32]

The human brain did not evolve primarily to discover truth. Like any other organ, it evolved to maximize the reproductive success of its owner. To find food, you need to have an accurate idea of what is edible and where it is to be found, and to protect yourself from predators, you need to be able to distinguish predators from other creatures accurately. However, in the social world, where deception and manipulation often rule, it sometimes pays not to know too much about your own agenda. In these circumstances, the person who is too insightful often ends up the loser.

7

MORAL PASSIONS

> He is the only one who remained true to himself, who did not cheaply sell his faith and his ideals, who always and without doubt followed his straight path toward his goal.
>
> —JOSEF GOEBBELS, SPEECH ON HITLER'S FIFTY-SIXTH BIRTHDAY

WAR IS A MORAL ISSUE. Arguably, it is *the* moral issue because it is difficult to envisage any activity that is of greater human consequence than war. From ancient times to the present, people have agonized over the question of whether war can ever be ethically justified, and there have always been conscientious objectors: men for whom war is so irredeemably base that they prefer opprobrium, prison, torture, and even death to participating in it. However, there is another, darker dimension of the relationship between war and morality. Aggressors are often *inspired* by moral feelings. They conceive of war primarily as a moral campaign and a religious, or quasi-religious, mission. Consider Adolf Hitler. In the minds of many people, he represents the very model of evil. But Hitler did not set out to do evil, but to deliver the world from depravity. "Theater, art, literature, cinema, press, posters and window displays," he righteously declared, "must be cleansed of all manifestations of our rotting world and placed in the service of a moral, political and cultural idea."[1] Strange as it may seem, the extermination of European Jewry represented the culmination of these deeply felt aspirations. According to historian Lucy Dawidowitz:

> The Jews inhabited Hitler's mind. He believed that they were the source of all evil, misfortune, and tragedy, the single factor that, like some inexorable law of nature, explained the workings of the universe. The irregularities of war and famine, financial distress and sudden death, defeat and sinfulness—all could be explained by the presence of that single factor in the universe, a miscreation that disturbed the world's steady ascent towards well-being, affluence, success, victory. A savior was needed to come forth and slay the loathsome monster. . . . The Jews were the demonic hosts whom he had been given a divine mission to destroy.[2]

The Khmer Rouge *purified* Cambodia of nearly two million human beings during the 1970s. It was, according to François Ponchaud, in his harrowing book *Cambodia: Year Zero*, "the translation into action of a particular vision of man: a person who has been spoiled by a corrupt regime cannot be reformed, he must be physically eliminated from the brotherhood of the pure."[3] War is almost always conceived as a battle between good and evil (*we* are good, *they* are evil) that is framed in religious or quasi-religious terms. The Puritan settlers of New England branded the native Pequot Indians "agents of Satan" and rejoiced in their death. After attacking and burning down a Pequot village containing five or six hundred noncombatants in 1637, Plymouth's governor, William Bradford, wrote that "it was a fearful sight to see them thus frying in the fire and the streams of blood quenching the same . . . but the victory seemed a sweet sacrifice, and they gave praise thereof to God, who had wrought so wonderfully for them . . . so speedy a victory."[4] Sometimes the righteous exhortations on opposing sides of a conflict sound eerily alike. George W. Bush proclaimed in his 2003 State of the Union address that "God told me to strike at Al Qaeda and I struck them. And then he instructed me to strike at Saddam, which I did. With the might of God on our side we will triumph." On the eve of the American invasion of Iraq, Saddam Hussein entreated, "O Arabs, O believers across the world, O enemies of evil, God is on your side. Rely on God and the soldiers of the Merciful on our land will be granted victory."[5]

What are we to make of the strangely equivocal relationship between morality and war? Why is it that moral passions so often spawn terror, slaughter, destruction, and oppression? To find answers we must delve into the evolutionary psychology of moral feeling.

WHAT IS MORALITY?

Morality is fundamental to our lives. But what exactly is it? Philosophers have batted around this question for thousands of years, and have produced a wide range of putative answers. One possibility is that good and evil exist objectively "out there" in the world, completely independent of subjective human judgment. This theory grants moral values a certain permanence and solidity by rending them impervious to the shifting currents of emotion and belief. It also implies that there is some fact of the matter about whether an act is good or evil, right or wrong, and that moral judgments are more than merely matters of opinion. Unfortunately for the moral objectivist, this theory is difficult to square with a scientific conception of human nature. To appreciate why, think for a moment about how we perceive the world. We use our senses. We know that a pineapple is sweet by tasting it, that a tomato is red by seeing it, and that a crow's caw is raucous by hearing it. We know that the gas tank is almost empty by looking at the gauge, and that the soup has a hundred calories per serving by reading the label. Our sense organs give us information about the size, texture, weight of objects, their color, their smell, and all of their other qualities. If moral qualities like good and evil are part of the objective furniture of the world, we must be able to perceive them in much the same way that we perceive other natural phenomena—by using our senses. But our sense organs do not give us any information about moral values. To bring this home, imagine a police inspector at the scene of a murder. Examining the corpse, she notices that it is the body of a bald man with blue eyes. He is around fifty years old, has a muscular build, and is slightly overweight. She sees that he is wearing a deep red sweater and a pair of blue jeans and that there is a bullet hole in his left temple. She can observe all of these facts, and more. But she cannot observe the moral badness of the act of murder. Whatever evil is, it is certainly not something that can be seen, heard, touched, tasted, or smelled. If good and evil were really objective features of the world, we would need some form of extrasensory perception to detect them. The idea that moral values are objective simply does not hold water.

Turning to works of literature can be helpful when trying to understand puzzling features of human nature, which is why generations of psychologists have plumbed the works of great playwrights and novelists. I have found Mark Twain's writings to be a particularly fertile source of ideas about the human condition. Although he is mainly remembered as a great American humorist, Twain was also an extraordinarily profound observer of human nature, and his works were devoured by the likes of Charles Darwin, Sigmund Freud, and William James.[6] His greatest novel, *The Adventures of Huckleberry Finn*, contains a wealth of insights into human psychology, including some very telling descriptions of moral deliberation. Huck, a backwoods boy, feigns his own death to escape from his brutal, alcoholic father. Jim, a runaway slave who is trying to reach the Free States, accompanies him up the Mississippi River on a makeshift raft. In the antebellum South, harboring an escaped slave was not only illegal, it was considered as gravely immoral, and Huck—the narrator of the story—is tormented by conscience.

> Jim talked out loud all the time while I was talking to myself. He was saying how the first thing he would do when he got to a free state he would go to saving up money and never spend a single cent, and when he got enough he would buy his wife . . . and then they would both work to buy the two

> children, and if their master wouldn't sell them, they'd get an Ab'litionist to go and steal them. . . . It most froze me to hear such talk. He wouldn't ever dared to talk such talk in his life before. Just see what a difference it made in him the minute he judged he was about free. . . . Coming right out flat-footed and saying he would steal his children—children that belonged to a man I didn't even know; a man that hadn't ever done me no harm. . . . My conscience got to stirring me up hotter than ever, until at last I says to it, "Let up on me—it ain't too late, yet—I'll paddle ashore at the first light, and tell." I felt easy, and happy, and light as a feather, right off. All my troubles was gone.

Not long after these deliberations, Huck rows ahead in a canoe and two armed men hail him from their skiff. The men are bounty hunters tracking down fugitive slaves. They realize that Huck has a companion, and ask whether he is white or black. Huck is faced with a dilemma.

> I didn't answer up prompt. I tried to, but the words wouldn't come. I tried, for a second or two, to brace up and out with it, but I warn't man enough—hadn't the spunk of a rabbit. I see I was weakening; so I just give up trying, and up and say . . . "He's white."[7]

Many philosophers, from Plato to the present, have claimed that moral dilemmas are a struggle between reason and passion, and that morality triumphs only when reason gets the upper hand and pins passion to the mat. But Twain's account tells a different story. His protagonist does not make a *rational* decision about the best thing to do. He does not experience a battle between cool reason and hot passion. Instead, his agonized deliberations are a matter of *conflicting passions* vying for the control of his behavior. On Twain's account, it is emotion, not reason, that lies at the heart of morality.

This thesis was set out more formally almost a century before

Twain's birth by David Hume, who scandalized his contemporaries by arguing that morality is always a matter of passion or feeling and that reason is always a slave to the passions. Think again about the law-enforcement officer at the murder scene. She knows that the murder is evil, but she cannot *perceive* its wickedness. Where, then, is it to be found? Hume had an answer. "You can never find it," he wrote, "till you turn your reflexion into your own breast. . . . It lies in yourself, not in the object."

> An action, or sentiment, or character is virtuous or vicious; why? Because its view causes a pleasure or uneasiness of a particular kind. In giving a reason for the pleasure or uneasiness, we sufficiently explain the vice or virtue. To have the sense of virtue is nothing but to *feel* a satisfaction of a particular kind. . . . The very feeling constitutes our praise or admiration.[8]

Now, over two hundred fifty years later, Hume's insights have begun to garner scientific support. Some of this comes from the work of Harvard University psychologist Marc Hauser. Hauser asks his research subjects questions about imaginary moral dilemmas such as the devilish scenario called the "trolley problem" devised decades ago by the British philosopher Philippa Foot. Here's how it goes: A man out taking a stroll notices a speeding, out-of-control trolley hurtling toward five people who are walking at a leisurely pace along a stretch of empty track. Their backs are toward the trolley, and they are blithely unaware of the mortal danger bearing down on them. By quickly pulling a lever, the onlooker can divert the trolley from its path and save the lives of the five pedestrians. But if he does this, the diverted trolley will run over a lone person who is standing on an adjacent stretch of track. What should he do? What would *you* do? The picture gets even more interesting if we fiddle with the parameters. What if there is no lever with which to stop the trolley. However, a very fat man happens to be standing nearby at the very moment you realize the danger the five pedestrians are in. Sizing up the situation, you see that if you shove him onto the track, his massive body will stop the trolley. Would you kill one innocent person to save the lives of five? Hauser found that most people—a whopping 89 percent of them—opt for pulling the switch in the first scenario, but the same people recoil at the prospect of pushing a man to his death, even though in both situations one life is sacrificed to save five. Even more interestingly, although most people make the same choices, Hauser found that they give suspiciously varied and often patently inadequate justifications of their decisions. He concluded from this that people make moral decisions unconsciously. They get a "gut feeling" for which they subsequently concoct seemingly rational (or not so rational) justifications. Sometimes, we can't even manage to scrape together a rationalization, and are tongue-tied when asked precisely why we consider an act to be right or wrong. When we make moral judgments "the emotional dog wags its rational tail."[9]

Hume's general account of morality seems to be vindicated by these experiments. But recognizing that morality is a matter of feeling is only a first step. We need a detailed explanation of *why* we respond to some situations with feelings of approval and others with disapproval. As usual, Hume anticipated this question and offered a brilliant answer. He argued that moral feelings are underpinned by "sympathy," a state that we would nowadays call compassion—literally *suffering with* another person. Something about human nature causes us to resonate with the people around us. We can even feel sympathy for purely fictional beings, which is what makes reading novels such fun (imagine how boring it would be if you simply didn't care what happened to the protagonist).

Morality does not have to be drilled into us—in fact, it cannot be. Moral feelings emerge spontaneously. Hume, who was a naturalist to the core, would be pleased to know that in the early 1990s an Italian neuroscientist named Giacomo Rizzolatti discovered neurons in the brains of his experimental monkeys that gave empirical substance to these speculations. These neurons had a very peculiar property. They became active both when a monkey performed an action and *also* when it merely *observed* another monkey performing the same action.

Rizzolatti decided to call them "mirror neurons." Further research revealed that mirror neurons are found in the brains of human beings, and that when we observe someone performing an action, our brains automatically respond as though we were performing the action ourselves. Thanks to motor neurons, a couch potato sitting in front of the TV watching a miniature, two-dimensional image of Alex Rodriguez hitting a home run feels as though he slamming the ball along with him and everyone at a Rolling Stones concert *becomes* Mick Jagger strutting across the stage. But there is more. The mirror neurons in human brains don't just simulate other people's actions. They are specialized to detect other people's *psychological states.* "Mirror neurons allow us to grasp the minds of others," says Rizzolatti, "not through conceptual reasoning but through direct simulation. By feeling, not by thinking."[10] Rizzolatti and his colleagues do not go as far as demonstrating that morality is underwritten by mirror neurons, but they come pretty close.

The notion that morality is based on compassion sounds wonderful in theory, but even a quick glance at the state of the world would seem to justify an opposite conclusion. Most people are indifferent to the bulk of human suffering. I doubt that many readers of this book lose any sleep about the hunger, disease, and violence that millions of human beings experience every blighted day of their lives. Most are probably far more concerned about matters like losing that extra ten pounds, getting a promotion, or organizing their next vacation than they are about the millions of men, women, and children whose bodies are being ripped apart in foreign wars or slowly ravaged by malnutrition and disease. The impressive record of atrocities racked up by the human race does not suggest that our conduct is guided by sympathy for others. Unfettered compassion would make saints of us all, but as Hume was well aware, compassion is not unfettered. It comes with strings attached. Our feelings of sympathy do not embrace all of humanity in equal measure. Some human beings *matter* to us. We care intensely about their well-being. Others do not matter very much, and still others do not matter at all. This is a hard saying, and may be difficult to accept but it is obviously and undeniably true.

Why is this? Hume's answer was characteristically insightful, and anticipated by more than two hundred years the discoveries of contemporary social psychologists. He argued that our sympathies are skewed by three biases. The first is a bias toward similarity. We favor people who resemble us—who look like us, dress like us, speak our language, and share our beliefs and skin color. Speaking bluntly, we feel that people who resemble us are more valuable than those who do not. To many Americans, the attack by Al Qaeda on the World Trade Center was a unique and unparalleled tragedy. Objectively speaking, this is not really true. The 9/11 fatalities were slight compared to the number of lives lost in the ongoing genocide in Darfur, the civilian deaths caused by the war in Iraq, and the devastation wrought by violence, malnutrition, and disease in the Democratic Republic of the Congo. We Americans do not care much about the fate of these strange, foreign people. Their suffering elicits little sympathy. As Chris Hedges remarks:

> While we venerate and mourn our own dead we are curiously indifferent about those we kill. Thus killing is done in our name, killing that concerns us little, while those who kill our own are seen as having crawled out of the deepest recesses of the earth, lacking our own humanity and goodness. Our dead. Their dead. They are not the same.[11]

Second, we are biased toward those with whom we come into direct contact. Moral feelings conform to the adage "Out of sight, out of mind": an injustice that unfolds in front of our eyes evokes passionate outrage, but one that occurs far away and affects only strangers leaves us cold. This is why the sight of a weeping child evokes a greater flood of feeling than the knowledge that that there are now approximately 11 million AIDS orphans in sub-Saharan Africa, many of whom are dying. (If you didn't know that before, you do now. Try a little experiment and note to what degree and for how long this piece of information troubles you. I doubt that it will do so for very long.) Christ's injunction to "love your neighbor" is, in the end, not all that difficult to satisfy. It is a much greater challenge to love someone who is *not* your neighbor, someone who lives on the other side of town or on the other side of the world.[12] Hume's third bias is based on kinship. We are moral nepotists, favoring family members over people who are unrelated to us. The lives of our relatives are felt to be more valuable than the lives of others, and their sufferings seem to be of greater consequence. Imagine the following scene: A house is on fire. Your only child is locked in one bedroom, and three children unrelated to you are locked in another. There is just enough time to rescue the occupants of *one* of the bedrooms. What should you do? If you are like most people, you would unhesitatingly choose to save your own child and let the other three burn. In fact, a person who, being more altruistic, would let his child die to save the lives of the three others would probably be regarded by most people as a psychopathic monster.

Hume's theory suggests that it is natural for human beings to be ethnocentric, xenophobic, and nepotistic. This is why he thought that we should try to overcome these built-in biases, and to treat other people more evenhandedly. Hume thought that we need to make a conscious effort to adopt what he called "the general point of view." This is based on the idea that moral principles must apply across the board, and therefore "what is wrong for you cannot be right for me merely because I am I and you are you."[13] An ideally moral person would treat *all* people with equal respect and compassion. Unfortunately, human nature does not always play ball with our noble ideals; consequently, we slip into racism, nationalism, and other forms of bigotry all too easily. However well intentioned, the "general point of view" turns out to be a pathetically fragile dam against the mighty tide of passion and prejudice that flows through human affairs.

THE BIOLOGY OF BIAS

Hume's theory dovetails remarkably well with an evolutionary account of human behavior. His triad of biases appears to be underwritten by nature herself, etched into the structure of our brains by natural selection. As one of the first thinkers to reject the artificial barriers that we so arrogantly erect between nonhuman animals and ourselves, Hume would probably have been delighted by this vindication of his theories.

Sociobiology teaches us that the moral prejudice is bound up with the evolution of altruism. Biologists use the term "altruism" for any behavior that promotes the reproductive success of another at one's own expense. For example, the black racer (*Coluber constrictor*) is a harmless snake native to the eastern United States. Female racers lay their eggs in early summer and guard them zealously. They have even been known to attack human beings that come too near their nests. Fully grown black racers are only around five feet long. They are not venomous, and their small, fragile teeth leave nothing more damaging than two rows of pinpricklike punctures in the flesh of a would-be assailant. Human beings are much larger and stronger than these snakes and can easily kill them. Any black racer foolhardy enough to assault a human being is therefore risking her life to protect her unborn young, endangering herself for their benefit. This behavior is not unique in nature. There are many other organisms that compromise their well-being and even give their lives for the benefit of others. Our own species presents many examples—from everyday, small-scale acts of generosity and kindness to heroic self-sacrifice.

Altruistic behavior may seem puzzling because the theory of evolution *seems* to imply that only the most obsessively self-interested organisms would survive the relentless meat grinder of natural selection. After all, every calorie expanded to benefit someone else's reproductive agenda is a calorie not spent furthering one's own. Or is it? In the early 1960s a British biologist named Bill Hamilton realized that explaining altruism required a complete overhaul of the conventional notion of reproductive success. Normally, what is good for you is also good for your genes—so when looking after your interests, you are also looking after theirs. This is true because you are a package for your genes, the "vehicle" in which they travel through space and time.[14] But copies of your genes also come in other packages, and the greatest concentrations of them are wrapped in the flesh of your closest relatives.

Each of your parents, siblings, and children carry, on average, 50 percent of your genes. Hamilton elegantly demonstrated that if we stop concentrating on whole organisms, and conceive of reproductive success as the proliferation of *genes*, it is easy to understand the Darwinian logic that inspires the behavior of animals that make sacrifices for their kin. Making babies is just one way of promoting the spread of your genes. Helping your relatives make babies is another path to the same destination. So, beefing up the reproductive prospects of your closest relatives is just another way of ensuring the success of your genes.

This was fine, as far as it went, but Hamilton's theory was silent on the important matter of altruism between nonrelatives. Eight years after Hamilton's discovery, sociobiologist Robert Trivers—then at Harvard University—demonstrated that it pays to be generous to your neighbor if your neighbor can be counted on to be generous in return. As long as the costs do not exceed the benefits, natural selection favors organisms that share with one another. Trivers called this you-scratch-my-back-I'll-scratch-yours attitude "reciprocal altruism," to distinguish it from the blood-is-thicker-than-water "kin altruism" explained by Hamilton.

Trivers also spoke of "generalized altruism," a notion that was further developed by University of Michigan biologist Richard D. Alexander in his concept of "indirect reciprocity." Reciprocal altruism often involves a direct exchange: the individual who receives a benefit eventually repays it to the donor. However, many altruistic acts are not at all like this. When you give money to a panhandler, the likelihood of being recompensed by the beneficiary of your charity is astronomically remote. Cases like this led Alexander to formulate the idea of indirect reciprocity. Indirectly reciprocal acts are beneficial to the agent, but in a roundabout fashion. "Cast your bread upon the waters," the Bible tells us, "for after many days you will find it again," or, more colloquially, "what goes around comes around." Obvious displays of largess benefit the altruist by boosting his or her reputation in the community; the altruist is seen as a "good" person, someone who can be trusted and with whom it is beneficial to have dealings.[15]

You have probably noticed that Hume's three principles fit biological theory like a glove. Humean nepotism is identical to Hamilton's "kin altruism": we favor family members because they share our genes. Hume's observation that we are biased toward people with whom we have direct contact can be explained by the fact that, as members of the same community, we are enmeshed with them in a web of interdependent, reciprocally altruistic relationships with them. Finally, Hume's point that we are partial to people who resemble us in behavior or appearance makes excellent biological sense given that these individuals are likely to share our genes or be members of our community or both. It is remarkable that, over a hundred years before Darwin wrote *The Origin of Species*, and two hundred years before the seminal contributions by Hamilton, Trivers, and Alexander, a philosopher with no knowledge evolutionary of biology could get things so right.

THE SIXTH COMMANDMENT

Earlier in this book I argued that war has probably been a feature of human life from the beginning, and I presented some facts and figures to demonstrate how widespread it is. Now it is time to consider the opposite side of human nature. Human beings have powerful inhibitions against killing one another.

Human groups are very dangerous to one another. But there another side to this story. We are an extremely social species, and it is important to bear in mind that our ancestors triumphed not as individuals, but as members of victorious *communities*. To accomplish this, they needed to maintain a very high level of cohesion and solidarity, which in turn required powerful barriers against in-group violence. The principle is a simple one: if community members are busy killing one another, they cannot present a united front against an enemy. The biologist Robert Bigelow was perhaps the first person to fully recognize the dynamic interplay between in-group cohesion and intergroup violence. "We are without doubt," he remarked, "the most cooperative and the most ferocious animals that ever inhabited the earth."[16] We cooperate to compete, and a high level of fellow feeling makes us better able to unite to destroy outsiders.[17] Bigelow sardonically commented:

> A hydrogen bomb is an example of mankind's enormous capacity for friendly cooperation. Its construction requires an intricate network of human teams, all working with single-minded devotion towards a common goal. Let us pause and savor the glow of self-congratulation we deserve for belonging to such an intelligent and social species. Without this high level of cooperation no hydrogen bomb could be built. Without an equally high potential for ferocity, no hydrogen bomb ever *would* be built. Perhaps our cooperation has something to do with our ferocity.[18]

Even the way that we play testifies to this. Many juvenile mammals play rough and tumble games, but nonhuman competitive play almost always pits one *individual* against another, as when two kittens "fight." A whole group of chimpanzee youngsters will sometimes play against a single individual (as human children do in games like blindman's buff). But only human beings play games in which *teams* try to defeat one another, a pattern found both in the playground and in sporting competitions witnessed by millions of fans. All such games have players and observers, which raises the question of why we are so fascinated by mock combat between cooperative groups. Play is a rehearsal for life. Kittens play-fight one-on-one in preparation for their territorial squabbles as adults, and when juvenile chimpanzees play many-against-one they are preparing for the raids that they will undertake as adults. When human beings play team sports, or watch them (remember those mirror neurons!), they are unconsciously rehearsing for war. Just look at team sports like football. The teams are organized like military units, they have territorial affiliations and distinctive uniforms and embody the martial virtues of strength, speed, and marksmanship. All the members of the team pull together to defeat a common opponent, and, like warriors of old, contemporary sports heroes are notoriously attractive to women.[19]

The dual heritage of cooperation between insiders and hostility against outsiders imbues our most cherished cultural and religious traditions. The sixth commandment, *lo tirtzach,* which is often mistranslated as "Thou shalt not kill," actually means "Thou shalt not murder"—more than a trivial difference, given that in the Old Testament God often commands the Children of Israel to kill. As I mentioned in chapter 3, God instructed His chosen people to exterminate the inhabitants of Canaan. He also demanded blood as punishment for disobedience. A man named Achan who, during the sack of Jericho (in which the Israelites "destroyed with the sword every living thing in it—men and women, young and old, cattle, sheep and donkeys"[20]), kept a little booty for himself instead of surrendering it to the Lord's treasury. In doing this, Achan violated one of the rules of *herem,* of holy war. His punishment was severe.

> Then Joshua, together with all Israel, took Achan son of Zerah . . . his sons and daughters, his cattle, donkeys and sheep, his tent and all that he had, to the Valley of Achor. Joshua said, "Why have you brought this trouble on us? The LORD will bring trouble on you today." Then all Israel stoned him, and after they had stoned the rest, they burned them.[21]

The Bible does not *forbid* killing: it restricts it. It is acceptable to kill strangers in war and to execute lawbreakers within the community: in the first instance the group cooperates to kill, and in the second it kills to cooperate. The fact that killing is rarely forbidden *absolutely* means that we have had to cultivate the ability to curb violence against other community members while maintaining the capacity to unleash it against outsiders. We are equipped with cognitive mechanisms that enable us to shift from one state to the other. This is what makes it possible for a loving husband and father to say good-bye to his family, go off to butcher other human beings, and eventually return, if he is fortunate, to resume his life as a law-abiding member of his society.

THE SYMBOLIC SPECIES

Do you remember the bacterium *Pseudomonas aeruginosa* described in chapter 2? The aggressive behavior of *Pseudomonas* is under the control of a simple environmental stimulus. The bacterium turns on its host in response to concentrations of interferon-gamma. There is a world of difference between this and the behavior of human beings. Unlike the bacterium, our behavior is not caused by raw environmental stimuli, but by our ideas, fantasies, hunches, and hypotheses about the meaning of those stimuli. The cognitive powerhouse of the human brain is incomparably more impressive than that of any other animal species. One of the most important features of this very special organ is its capacity to bring a wide range of concepts to bear on the world in which we live. Our brains also grant us the unique ability to indulge in fantasy and creative imagination, pretending and supposing, and to evaluate these brainchildren, distinguishing, at least in principle, good ideas from bad ones.[22]

The distinctive character of human warfare owes a great deal to these factors. Chimpanzees recognize and attack members of neighboring communities, but their behavior is never filtered through a web of beliefs about good, evil, pride, humiliation, friends, heroes, villains, and martyrs. No chimpanzee can dream of establishing a master race, of conquering the Holy Land, of seizing nonexistent weapons of mass destruction, or of undertaking a kamikaze mission in exchange for an eternity in paradise. There is no nonhuman equivalent of notions like manifest destiny—"the right . . . to overspread and to possess the whole of the continent which Providence has given us for the development of the great experiment of liberty and federative development of self-government entrusted to us," for there is nothing in the cognitive repertoire of any nonhuman species remotely like "right," "Providence," or "liberty."[23] The evolution of language initiated a quantum leap in the complexity of human behavior. We became diabolically symbolic creatures, whose culture-generating brains are a breeding ground for ideas with the power to kill. Only a diabolically symbolic creature could enact scenes like those burnt into the memory of a child who awoke one morning to find the crucified, beheaded, and castrated body of his father lying in front of his house with horseshoes nailed to his bare feet. Later that day he stood in a crowd in the town square watching Armenian women perform a dance of death for the amusement of a group of Turkish soldiers.

> They had whips and each had a gun. They were shouting "Dance. *Giaur, Slut.*" . . . Their hair had come undone and their faces were wrapped up in the blood-stuck tangles of hair, so they looked like corpses of Medusa. Their clothes were now turning red. Some of them were half naked, others tried to hold their clothes together. They began to fall down and when they did they were whipped until they stood and continued their dance. Each crack of the whip, and more of their clothing came off. . . . Then two soldiers pushed through the crowds swinging wooden buckets and began to douse the women with the fluid in the buckets, and, in a second, I could smell that it was kerosene. And the women screamed because the kerosene was burning their lacerations and cuts. Another soldier came forward with a torch and lit each woman by the hair. At first, all I could see was smoke, and the smell grew sickening, and then I could see the fire growing off the women's bodies, and their screaming became unbearable. The children were being whipped now, furiously, as if the sight of the burning mothers had excited the soldiers, and they admonished the children to clap, "faster, faster, faster," telling them that if they stopped they would be lit on fire.[24]

Scenes like this are so extravagant and bizarre, and so incredibly cruel, that one is tempted to simply label them as evil. But this would be vacuous: a condemnation, not an explanation. It is just another way of saying that we abhor it. But tell me why, if we find such acts so abhorrent, we repeat them again and again? There is clearly something ambiguous about our relationship with war and its attendant atrocities, something deeply contradictory and self-deceptive. Something that we seem to be unable or unwilling to grasp.

8
RELUCTANT KILLERS

My men, my modern Christs,
Your bloody agony confronts the world.
—HERBERT READ, *MY COMPANY*

HAVE YOU EVER CONSIDERED KILLING SOMEONE? Perhaps it was a rival, your boss, or a lover who spurned you? If you have never actually considered it, have you, perhaps, *imagined* it? And did the fantasy of putting a bullet through that hated person's head or slipping rat poison into their coffee give you a frisson of pleasure? If so, you are not alone. Most people—around 84 percent of women and 91 percent of men—admit to daydreaming about killing people whom they dislike. We don't just have private fantasies either; we indulge in collective ones. Our taste in film (prepackaged fantasies for the millions) often veers towards the warlike and murderous. Before there were movies, there was literature, and here, too, we often gravitate toward violent themes. The Greeks had the *Iliad*, the Elizabethans had *Macbeth* and now, as I write this paragraph, nine of the top ten bestselling fiction hardback books feature the theme of murder.

All of this suggests that there may be a homicidal streak in human nature.[1] However, only a minute percentage of people who daydream about murder go on to commit it. Rates of homicide are reassuringly modest. In 2004 only 5.5 out of every 100,000 Americans were victims, and this paltry figure represents the *highest* homicide rate in the developed world. This implies that less than .005 percent of the American population commit homicide (as some murderers kill more than once, there must be fewer killers than victims). Even in Jamaica, which with 63 incidents for every 100,000 people has the highest murder rate in the world, less then .06 percent of the population are killers.[2]

Why are murder rates so low? Perhaps our violent, antisocial impulses are restrained mainly by threat of punishment. This thesis was championed by Hobbes, who believed that in the absence of a "common power" to keep us "all in awe" the inevitable result is a "warre of all against all." It was portrayed by William Golding in *The Lord of the Flies,* which tells the story of a group of English schoolboys marooned on a remote island who gradually spiral downward into tribalism, brutality, and murder. Hobbes certainly had a handle on the truth. One need only look at the way that all Hell breaks loose when social infrastructures collapse and there is no effective authority to maintain order, as we saw when Hurricane Katrina roared into New Orleans in 2005. Although Hobbes was definitely on to something, he was wrong to think that the fear of punishment is the only thing that keeps us from one another's throats. If punishment is a deterrent to murder, then the supreme punishment, execution, should act as its ultimate prophylactic. But there is no evidence that capital punishment discourages murder. Seven of the ten countries with the *lowest* murder rates in the world do *not* allow capital punishment, and in the United States, the ten states with the *highest* homicide rates permit capital punishment, while six of the ten states with the *lowest* homicide rates forbid it. Of course, this is far from conclusive. It may be that the countries and states with elevated murder rates would be even worse without the death penalty. But it does show that there is something wrong about laying our reluctance to kill entirely at the doorstep of the fear of punishment.

This becomes even more apparent in situations where killing is encouraged. Military combat *requires* men and women to kill. In war, there is no question of being punished for taking other people's lives, so if it is true that the fear of punishment is all that stands between us and mayhem we would expect soldiers to be trigger-happy killing machines. But this is rarely the case. Special training is required to get servicemen to kill on demand, and even then psychological obstacles often stymie it. Real war is nothing like the heroic versions presented in motion pictures and adventure novels: most recruits find it very difficult to bring themselves to kill on the battlefield. The main spokesman for this idea was United States Army general S. L. A. "Slam" Marshall. Marshall's research, which was based on interviews with U.S. soldiers during World War II and the Korean War, led him to the astonishing conclusion that 75 to 85 percent of American soldiers refrain from firing their weapons in battle. As he put it in his classic work *Men Against Fire:*

> The average and healthy individual—the man who can endure the mental and physical stresses of combat—still has such an inner and usually unrealized resistance towards killing a fellow man that he will not of his own volition take life if it is possible to turn away from that responsibility.[3]

Marshall is a controversial figure, and his research has been savaged for its lack of methodological discipline (in fact, he has been accused of fabricating his figures). But this sort of pedantic objection is ultimately beside the point. It does not really matter that Marshall never provided empirical justification for his figures for ratios of fire. His general point, that soldiers are reluctant to kill, is supported by anecdotal evidence from many sources. During World War II, less than 5 percent of U.S. fighter pilots were responsible for 30 to 40 percent of the kills, and most pilots did not shoot down any planes at all.[4] Even soldiers at the Battle of Gettysburg—a truly horrendous engagement that involved a combined force of over 160,000 with some 7,000 killed in action—often seem to have avoided firing their weapons. After the fighting was over, 27,574 abandoned muskets were found on the battlefield, over 90 percent of which were loaded. The loading time for these weapons was nineteen times longer than the firing time, which means that only 5 percent of the guns *should* have been found loaded—that is, on the assumption that most of the soldiers were firing their weapons. "The only rational conclusion," writes military historian Gwynne Dyer, "is that huge numbers of soldiers at Gettysburg, both Union and Confederate, were refusing to fire their weapons even in the stand-up, face-to-face combat at short range, and were presumably going through the act of loading and perhaps even mimicking the act of firing when somebody nearby actually did fire in order to hide their internal defection from the killing process. And very many of those who did fire were probably aiming high."[5] Almost a century later, during World War II, Lieutenant Colonel Robert C. Cole of the 502nd Parachute Infantry complained that when his men were being attacked by German troops on the Carenton Causeway in Normandy, "Not one man in twenty-five voluntarily used his weapon. . . . I walked up and down the line yelling, 'God damn it! Start shooting!' But it did very little good."[6] One infantry platoon leader recounted, "Squad leaders and platoon sergeants had to move up and down the firing line kicking men to get them to fire."[7] Lieutenant Colonel Dave Grossman, a vociferous defender of Marshall's thesis, has accumulated many corroborating examples from the military literature, interviews, and correspondence with veterans and presents evidence going back to the early nineteenth century that soldiers often fired above their enemy's heads rather than *at* them. Marshall's research prompted the U.S. military to overhaul its training methods, which is apparently why a much higher firing rate was achieved in the Vietnam War and after.[8]

Although it is true that men have, from time immemorial, pursued carnage with bloodthirsty enthusiasm, it is also true men have probably always tried to avoid it. As social critic Barbara Ehrenreich notes in her book *Blood Rites:*

> Throughout history, individual men have gone to near-suicidal lengths to avoid participating in wars. Men have fled their homelands, served lengthy prison terms, hacked off limbs, shot off feet or index fingers, feigned illness or

> insanity, or, if they could afford to, paid surrogates to fight in their stead.[9]

Rates of desertion have often been incredibly high. To give just one dramatic example, in 1866–67, a full 52 percent of Custer's Seventh cavalry, and 25 percent of the entire frontier cavalry deserted! European officers used to avoid camping near woodland for fear that soldiers might slip away among the trees.[10] The coexistence in the human heart of the lust for war and the dread of it is central to the remainder of this book.

Although we tend to emphasize the fear of being killed or injured, the fear of killing also plays an extremely important role in the desire to escape going to war. Killing at close range is one of the most traumatic aspects of combat, and many men will do almost anything to avoid it. "Intimate brutality," as Grossman calls it, can have powerful and extremely disturbing psychological consequences.[11] In one typical account, a soldier described how, after bayoneting a German infantryman at the Battle of the Somme, he "started to shake and . . . shook like a leaf on a tree for the rest of the night." A Vietnam vet who admitted to participating in massacres of women and children would "vomit for hours" after each mission.[12] One of the most evocative descriptions of a kill is found in William Manchester's powerful memoir *Goodbye Darkness.* Manchester cornered a Japanese soldier who was trapped in his own sniper's harness, and therefore unable to defend himself. Manchester killed him, and then continued pumping bullets into the corpse ("wasting government property," as he put it). The soldier fell to the ground, his eyes glazed over, and flies, which are ubiquitous on the battlefield, began to gather on his eyeballs. "I don't know how long I stood there staring," he wrote. "A feeling of disgust and self-hatred clotted darkly in my throat, gagging me."

> Jerking my head to shake off the stupor, I slipped a new fully loaded magazine into the butt of my .45. Then I began to tremble, and next to shake, all over. I sobbed, in a voice still grainy with fear: "I'm sorry." Then I threw up all

> over myself. I recognized the half-digested C-ration beans dribbling down my front, smelled the vomit above the cordite. At the same time I noticed another odor; I had urinated in my skivvies. . . . I knew I had become a thing of tears and twitchings and dirtied pants. I remember wondering dumbly: "Is that what they mean by conspicuous gallantry?"[13]

War does violence to the warrior, for in addition to the extremes of terror and fatigue confronting soldiers, they must also find some way of coming to terms with the enormous guilt that arises with the taking of human life. In the popular imagination, soldiers unflinchingly perform their duties and are emotionally unscathed by their experiences. But this image is tragically flawed. It is a convenient fiction that occludes a more disturbing reality. Real soldiers often experience physical symptoms of stress like violent palpitations of the heart, a condition that is so common in combat that nineteenth-century military physicians named it "soldier's heart." Many have episodes of uncontrollable trembling ("war tremors") which can be so violent that they are unable to use their weapons. They may even cause them to collapse in convulsions, literal paroxysms of fear. When this happened to some of Major Reno's men during the Battle of the Little Big Horn, one convulsing cavalryman had to be tied down to keep him from injuring himself. They may feel faint, break out in cold sweats, or have recurring nightmares, and sometimes the nightmares recur for the rest of their lives. Shakespeare mentions this phenomenon in *Henry IV* when he has Lady Percy say:

> *In thy faint slumbers I by thee have watch'd,*
> *And heard thee murmur tales of iron wars. . . .*
> *Thy spirit within thee hath been so at war*
> *And thus hath so bestirr'd thee in thy sleep,*
> *That beads of sweat have stood upon thy brow*
> *Like bubbles in a late-disturbed stream;*
> *And in thy face strange motions have appear'd,*

Such as we see when men restrain their breath
On some great sudden hest.[14]

Soldiers may vomit from stress, or involuntarily urinate. In one survey of a U.S. infantry division in the South Pacific, over a quarter of the men said that they had vomited—not from illness but from anguish—and a fifth admitted to losing control of their bowels (in today's combat-stress classes, U.S. Marines are warned that 25 percent of them are likely to lose bladder or bowel control during battle). We know that these phenomena are not mere artifacts of modern warfare, because many of the same reactions are mentioned in documents from centuries past. In an Egyptian letter written three thousand years ago, an experienced veteran tells a rookie officer what it is like to go into battle, "Shuddering seizes you, the hair of your head stands on end, your soul lies in your hand." The ancient Greek historian Herodotus describes an Athenian soldier named Epizelus, who was struck blind after hallucinating a phantom warrior killing a comrade in arms. Plutarch mentions that Greek soldiers lost the ability to speak during the siege of Syracuse in 212 B.C.—they were literally struck dumb with terror. In the *Anglo-Saxon Chronicle*, which records the Saxon invasion and settlement of England, an account of a ninth-century battle between the Saxons and the Danes records that the Saxon chieftain vomited violently and repeatedly while advancing on the enemy.[15]

ALTERED STATES

How do soldiers deal with this? What is it that allows them to go on and to continue killing? Sometimes they resort to mind-altering drugs—a practice with an ancient military pedigree. Prehistoric warriors in central Russia guzzled down an extract of the hallucinogenic mushroom *Amanita muscaria* before going to war. The Vikings consumed the same hallucinogenic mushroom. Ancient Inca soldiers chewed coca leaves, from which we derive cocaine. The Scythian fighters of central Asia smoked cannabis. The Rajputs of India took opium to steady their nerves and prevent incontinence. Today, the Yanomami of the Amazon rain forest ingest hallucinogenic drugs, and the Masai of East Africa prepare themselves for war by drinking a potent concoction made from a variety of narcotic herbs.

Closer to home, Russian soldiers in both world wars took valerian, a mild sedative, and at least 10 percent of U.S. servicemen during World War II took amphetamines (a quarter of U.S. military prisoners were addicts). In El Salvador sedatives were so widely used that they were banned in garrison towns, and the widespread use of narcotics by U.S. soldiers in Vietnam is legendary.

However, of all the drugs that have ever been used by soldiers to help them endure the strain of battle, alcohol takes pride of place. Ancient Greek infantrymen drank wine before engaging with the enemy, as did the ancient Chinese. Fourteenth-century French knights got drunk on the eve of the Battle of Agincourt, as did Harold's knights on the eve of the Battle of Hastings. Aztec warriors drank pulque, an alcoholic drink made from the fermented juice of the maguey plant, to dull their emotions and release their inhibitions. In fact, the expression "Dutch courage" comes from the practice of English soldiers in the Netherlands who prepared for battle by drinking *genever*, a form of gin. British soldiers received a rum ration, which was issued just before battle. According to one British veteran of World War II, "We simply kept going on rum. . . . Eventually it became unthinkable to go into action without it." Members of the Nazi mobile killing squads—the *Einsatzgruppen*—were given schnapps to drink before and after mass killings, and the genocidal killers in Bosnia and Kosovo drank heavily, too. Soviet troops advancing through Berlin in 1945 fueled their rampage of rape and atrocity by drinking every intoxicant that they could lay their hands on, including toxic chemicals. Harvard University historian Niall Ferguson goes as far as to say that "without alcohol . . . the First World War could not have been fought. . . . Ordinary soldiers would get drunk at every opportunity." So did their officers. The poet and novelist Robert Graves knew officers who drank as much as two bottles of whiskey a day. Cognac was freely dispensed to Italian troops at the front—they called it their "gasoline." When U.S. soldiers in World War II ran out of booze they made their own or drank cocktails of Aqua Velva aftershave mixed with fruit juice.[16] And in today's scientific laboratories, researchers are trying to develop new drugs that will allow soldiers to kill without guilt.[17]

MELTDOWN

All of this is so contrary to the popular image of the hero that most people seem to tacitly assume that serious psychiatric casualties are quite rare. In fact, they are common. During World War II over a million American soldiers suffered from diagnosable and seriously debilitating psychiatric symptoms. Professor of politics and army intelligence officer Richard A. Gabriel informs us:

> In the US ground forces alone, 504,000 men were permanently lost to the fighting effort for psychiatric reasons—enough manpower to outfit 50 combat divisions! Of these, 330,000 men were lost to ground combat units in the European Theater and received separations for psychiatric reasons. Another 596,000 were lost to the fighting effort for weeks or months and eventually returned to the line, and still another 464,500 reported to medical facilities for [psychiatric] treatment without being admitted and were returned to the line almost immediately.

These are large figures, but they become all the more striking when Gabriel points out that "in four years of war, no more than about 800,000 U.S. ground soldiers saw direct combat. Of these, 37.5 percent became such serious psychiatric cases that they were lost to the military effort for the duration of the war." At one point, psychiatric casualties were being discharged more rapidly than new recruits were coming it. Over 24 percent of U.S. soldiers in the Korean War became serious psychiatric casualties, and the most recent studies show that almost 20 percent of Vietnam War veterans developed diagnosable psychiatric disorders and 9 percent still suffered from them over a decade after the war. These astonishing figures pertain to psychiatric symptoms experienced *during* a tour of duty, but not those that appear after demobilization. Adding these can make a big difference. For example, although less than 13 percent of the American soldiers in Vietnam reported serious psychiatric symptoms, it has been estimated that somewhere between 18 and 54 percent of them experienced psychiatric disorders after returning home.[18]

These statistics probably underrepresent the true incidence of psychiatric casualties. They are probably the result of underreporting, as seeking help for psychological problems is not consistent with the tough macho image demanded by military culture. It is also a fact that although some psychiatric symptoms are blatant, and obvious for all to see, others can be private and insidious, and often escape diagnosis. We do our heroes a disservice when we express compassion for the wounds inflicted on their bodies but ignore those inflicted on their souls.

"Must you have battle in your heart forever? The bloody toil of combat?" sang Homer in *The Odyssey*, "That nightmare cannot die, being eternal evil itself—horror, and pain, and chaos."[19] For many veterans, the nightmare never ends. Audie Murphy, the most decorated soldier in American military history, was one such casualty. Murphy earned literally every known U.S. decoration for valor and personally killed over 240 enemy combatants during World War II. After the war, he became a well-known movie actor, and even played himself in a 1949 film adaptation of his autobiography *To Hell and Back*. But the all-American hero had a secret. He slept with a loaded revolver under his pillow. He would wake up from the violent nightmares that haunted his sleep and wildly fire his gun around the room "shattering mirrors, clocks and lightbulbs."[20] Murphy also suffered from depression, and drifted into compulsive gambling and womanizing. Another was T. E. Lawrence ("of Arabia"). Lawrence was the paradigmatic swashbuckling war hero. A leader in the Arab resistance against the Ottoman Turks, he became famous for his skill at guerrilla warfare. Lawrence began his military career with no discernible psychiatric disorders, but he emerged from it with a desperate sense of guilt and shame and a need to do penance by degrading himself. Craving humiliation, he even hired a man to beat him. Lawrence's friend Eric Kennington wrote that Lawrence wanted to "reclaim or re-create his soul"—a soul that he felt had been destroyed by war.

Given enough time, psychological damage is virtually inevitable in war. The U.S. Army concluded during World War II that most soldiers will experience psychiatric collapse after eighty to ninety combat days. Only around 2 percent of enlisted men do not break down no matter how long they have been subjected to the stress of war. You might think that these are exceptionally balanced, resilient, and emotionally mature individuals, but this is not their secret. These seemingly stable men usually have psychopathic personalities. They are unable to experience concern for others and enjoy the exercise of violence and cruelty. This is why, in the past, military psychologists recommended recruiting sociopaths and even men who had served time for murder or manslaughter. They believed that these dangerous men were ideal soldiers. Ironically, they often return home as decorated heroes. After sixty days of continuous fighting in Normandy, 98 *percent* of the survivors suffered psychiatric damage. ("The other 2 percent," remarks Dave Grossman, "were crazy before they got there.")[21]

It is natural to assume that the fear of injury and death are the cause of the psychiatric casualties of war and underestimate the significance of *the act of killing*. Soldiers and veterans often carry an immense burden of guilt because they may have done things in the course of duty that violate the primal taboo against killing one's own kind. These are not merely speculations: they are backed up by data. Despite the grave danger posed by war to civilians, noncombatants in war zones suffer far fewer psychiatric casualties than those who do the fighting. In a study of almost two thousand Vietnam vets, psychologist Rachel MacNair found that men who had killed in battle were significantly more likely to suffer from psychiatric symptoms than those who had not. To top it off, people who take human life without placing themselves in any danger (for example, public executioners) suffer from the same kind of psychological symptoms that military combatants do.[22]

In other societies, and indeed in our own in centuries past, it was understood that many soldiers bear enormous guilt for their actions. During the Middle Ages, for example, Christian warriors were required to do penance for a year or more to expiate the sins that they committed on the battlefield. In ancient Rome, Vestal virgins ritualistically bathed Roman soldiers to symbolically wash away the moral stain of killing. Similar ablutions were performed in societies as far-flung as the Masai of East Africa and the Plains Indians of North America.[23]

DISSOCIATION

Although frequently used, drugs have never been the main way that soldiers cope with war. They are merely adjuncts to the brain's internal resources. The human brain contains the psychological equivalent of circuit breakers that can mute or completely shut down the senses, and it can manufacture its own narcotics to make traumatic situations bearable. These are emergency reactions, and may have originally evolved to cope with attacks by predators. The nineteenth-century missionary David Livingstone's account of being mauled by a lion shows these internal mechanisms in action.

> He caught my shoulder as he sprang, and we both came to the ground below together. Growling horribly close to my ear, he shook me as a terrier does a rat. The shock produced a stupor similar to that which seems to be felt by a mouse after the first shake of the cat. It caused a sort of dreaminess in which there was no sense of pain nor feeling of terror. . . . It was like what patients partially under the influence of chloroform describe, who see the operation but feel not the knife.[24]

Soldiers often report that combat has a numbing, dissociative effect. They observe themselves without feeling that they inhabit themselves. Some, like U.S. Marine Philip Caputo, feel like they are watching a movie rather than having an experience. "I had enjoyed the killing of the Viet Cong." writes Caputo. "Strangest of all had been that sensation of watching myself in a movie. One part of me was doing something while the other part watched from a distance, shocked by the things it saw, yet powerless to stop them from happening."[25] Others take this one step further, and have the impression that their minds have been severed from their bodies. Compare Caputo's account with the following description by Ivone Kirkpatrick:

> Body and soul seemed to be entirely divorced, even to the extent that I felt that I no longer inhabited my body. . . . My mind was a distinct and separate entity. I seemed to hover at some height above my own body and to observe its doings and the doings of others with a sort of detached interest.[26]

Sometimes soldiers experience more extreme sensory distortions: they become blind, hallucinate, develop tunnel vision, and become deaf or hypersensitive to sound. The world may seem to crawl along in slow motion. Battle-weary soldiers may lose their sensitivity to pain and refuse treatment for wounds. Many slide into a chronic state of dissociation in which whole world feels unreal, a state signaled by the infamous thousand-yard stare—a weird faraway look in the eyes. There are many references to mental dissociation in the combat literature. For example, the German World War I veteran Ernst Jünger recounts how "we looked at all those dead with dislocated limbs, distorted faces, and the hideous colors of decay, as though we walked in a dream through a garden full of strange plants." T. E. Lawrence described how war "tore apart flesh and spirit," and many other veterans have spoken of their souls being "extinguished."[27] The dulling of the senses, the apparent severing of mind from body, and the feeling that the soul has been destroyed are automatic protective maneuvers.

War is not only an assault upon a soldier's body, it is an assault upon his sense of his own humanity. It requires him to perform acts that no human being should have to perform, exposes him to perpetual fear, and threatens him with crushing guilt. One way to minimize these pressures is to become distant, to close down, to enter into spiritual hibernation. When this happens, the world seems to become unreal. Joanna Bourke comments that:

> Instead of focussing on mangled corpses, soldiers who could imagine themselves as movie heroes felt themselves to be effective warriors. Such forms of disassociation were psychologically useful. By imagining themselves as participating in a fantasy, men could find a language which avoided facing the unspeakable horror not only of dying but of meting out death.[28]

Dissociation and drugs are not the only ways to make war bearable. To perform well in battle without succumbing to malaise, soldiers need a way to blunt the pain of warfare and overcome their natural horror of killing, while at the same time preserving or even enhancing their morale and effectiveness. This sounds like a very tall order, but evolution has endowed us with just this capacity. For this to happen, the soldier must immerse himself in a special form of self-deception. Strange as it may sound, his ability to deceive himself can make the difference between survival and extermination, victory and defeat. As advantageous as this is, it is also very dangerous, because it immobilizes his natural inhibitions against lethal violence and can unleash a torrent of brutality. My task in the next chapter will be to tell the story of how this occurs. This will take us to the heart of where war lives.

9

THE FACE OF WAR

> Anyone who has ever looked into the glazed eyes of a soldier dying on the battlefield will think hard before starting a war.
>
> —OTTO VON BISMARK, SPEECH, AUGUST 1867

WE ARE NOW IN A POSITION to bring the puzzle of war into sharp focus. The track record of our species shows, beyond a shadow of a doubt, that we are extremely dangerous animals, and the balance of evidence suggests that our taste for killing is not some sort of cultural artifact, but was bred into us over millions of years by natural and sexual selection. But we have also seen that there is something in human nature that recoils from killing and pulls us in the opposite direction. These contrary dispositions exist side by side within us, and any explanation of war must honor the tension between them. It is incorrect to claim, without qualification, that we are killer apes, but to say that we are essentially peaceable is every bit as misguided. We are *ambivalent* about killing, and it is impossible to understand the relationship between war and human nature without taking this into consideration. In this chapter I will begin to gather the strands of my argument together and show how several features of human nature—the love of killing, the dread of killing, self-deception, and the modularity of the mind—all conspire to make war possible.

IMAGINED COMMUNITIES

The remarkable similarities between chimpanzee and human warfare make it easy to lose sight of their equally striking differences. It is normal for chimpanzees to attack and kill others just because they belong to another community. Although we human beings are often hostile to strangers, we do not routinely hunt down and slaughter people simply because they live in a different locale than we do. On the other hand, human beings are capable of perpetrating acts of violence that are rare among chimpanzees, for we band together to butcher members of our *own* communities—people who live in the same place that we do and with whom we have dealings on a daily basis.[1]

These contrasting patterns of aggression are reflected in correspondingly distinct patterns of coalition. Amongst chimpanzees, social solidarity is entirely an in-group affair. Individuals may, and often do, form cliques to acquire status and power, but membership is restricted to members of one's own larger community. Furthermore, there is no such thing as a coalition *between* chimpanzee communities. In contrast, our species establishes coalitions between entire communities: at any given moment, some groups count as friends and others as enemies.

At some point in their evolutionary journey, our Stone Age ancestors moved from the chimpanzee pattern of intergroup violence to a distinctively human mode. If we can find an explanation for this, we will have learned something very important about human nature and about war.

The solution is simple, obvious, and revealing. It lies in the distinctive ways that chimpanzees and human beings form community groups and in the mental powers that account for these differences. Chimpanzee communities are local breeding populations that occupy specific geographical territories. When these groups split in two, as sometimes occurs, each group becomes a separate entity and has its own fiercely defended territory. When this happens, inhibitions against violence evaporate, as former comrades become legitimate targets for aggression. Primatologist Jane Goodall observed this when the chimpanzee community at Gombe (mentioned in chapter 4) divided into the Kahama and Kasakela groups. Apes who had lived peacefully together for years now became implacable enemies. "Often when I woke in the night," she recalls, "horrific pictures sprang unbidden to my mind. . . .

> Satan cupping his hand below Sniff's chin to drink the blood that welled from a great wound on his face; old Rodolf, usually so benign, standing upright to hurl a four-pound rock at Godi's prostrate body; Jomeo tearing a strip of skin from Dé's thigh; Figan charging and hitting again and again, the stricken, quivering body of Goliath, one of his childhood heroes. And, perhaps worst of all, Passion gorging on the flesh of Gilka's baby, her mouth smeared with blood like some grotesque vampire from the legends of childhood.[2]

Chimpanzees play by different rules than we do. From their perspective, past loyalties count for nothing: all that matters is whether or not an individual is a member of the same territorially defined breeding group. This is not a failure of memory, as chimpanzees have excellent memories and are able to recognize an individual after many years of separation.[3] Instead, it is based on how they draw their social boundaries: in-groups and out-groups are in or out of a relatively fixed physical territory. In contrast, humans draw their boundaries less rigidly and often live in widely dispersed groups: the Christian, Jewish, and Muslim communities, for example, have members who scattered all over the globe. Unlike chimpanzees, who interact on a more-or-less daily basis, most members of these large human communities never encounter one another at all. They are what Cornell University sociologist Benedict Anderson calls *imagined communities* that occupy conceptual rather than geographical space. Imagined communities are sustained by symbolic rather than genetic kinship. So, for example, internationalists talk about "international brotherhood," "the brotherhood of man," "the family of man"; participants in the Black Power movement called one another "brother" and "sister," as do members of Christian religious orders. Americans ask their nation to "crown thy good with brotherhood" and citizens of the former Soviet Union lifted their voices to "sing to our motherland, home of the free." Christians are all brothers and sisters, and the Koran proclaims that "the Believers are but a single brotherhood." The language of kinship permeates social discourse.[4]

ESSENCES AND ACCIDENTS

Human beings are able to create these communities because of a unique intellectual achievement: they are able to form *concepts*. Concepts are best understood as ways of classifying things and arranging them into groups. Although chimpanzees are highly intelligent, conceptual thought seems to be completely beyond them. Sure, they can recognize edible things, but there is no evidence that they have a *concept* of food, and they respond to one another as conspecifics, but there is no evidence that they have a *notion of what it is* to be a chimpanzee. Now, what is it that determines membership in one of these conceptual sets? That is a surprisingly difficult question to answer. Let's use the example of the concept "tiger." What is it that makes something a tiger? One (misguided, but very tempting) way to go with this question is to say something like "A tiger is an orange, black-striped, four-legged feline." But this is obviously inadequate, because it is possible for there to be albino tigers and three-legged tigers. But is it possible to have a nonfeline tiger? No. It seems that some characteristics—like the felineness of tigers—are especially important for determining what a thing is. Philosophers from Aristotle onward have tried to capture this distinction by the notions of *substance* (what a thing is) and *accidents* (qualities that a thing has). In this book, I will use the intuitively more appealing term "essence" for that something-or-other that makes a thing the kind of thing that it is, instead of sticking with "substance," which would be liable to confuse nonphilosophers. Crudely put, then, a tiger *is* feline (its essence) but *has* four legs, an orange coat, and black stripes (its accidents).

Philosophical debate has raged around these concepts for more than two thousand years and has led to some very strange ideas. According to Roman Catholic dogma, the Eucharistic host and sacramental wine do not merely symbolize the body and blood of Christ, they literally *are* the body and blood of Christ. The obvious problem with this is that they just don't look or taste like flesh and blood (if the transformation was an act of stage magic, it would not be very impressive). The Church got around this problem by the doctrine that the taste and appearance of the wafer and wine are mere "accidents," but the *substance* of the wafer and wine has been replaced by the *substance* of the flesh and blood of Christ. It follows that Holy Communion is a real rather than a metaphorical act of cannibalism! This bizarre idea caught on because it spoke to our natural tendency to distinguish between what something *really* is and what it merely presents itself as being. It is very easy for us to think that objects possess a mysterious "something" that makes them the kind of things that they are and that is quite independent of anything that we can observe about them. We naturally think of human beings in precisely this way: what makes a person a person is their possession of an intangible "soul" or "self," not their age, height, weight, hair color, or DNA. Thinking in this way quickly slides into dualism (I am a "soul" distinct from my body), and this leads to notions like life after death (my body rots, but my spirit is "incorruptible").

Essentialist thinking is found throughout the world and in all periods of recorded history. This way of thinking about things is so pervasive that, according to University of Michigan anthropologist Scott Atran, it may be an evolved feature of the human mind. Atran found that members of diverse cultures tend to classify living things in broadly similar ways, which suggest that there is an innate cognitive adaptation at work. He writes that:

> People in all cultures, it appears, consider this essence responsible for the organism's identity as a complex entity

> governed by dynamic internal processes that are lawful even when hidden. The essence maintains the organism's integrity even as it causes the organism to grow, change form, and reproduce. For example, a tadpole and a frog are conceptualized as the same animal even though they look and behave very differently and live in different places.[5]

Anthropologist Francisco Gil-White takes this idea farther. Not only do we automatically conceive of living things as members of species imbued with invisible "essences" that make them what they are, our brains also treat human groups in the same way. We see them as species—as entities possessing a hidden essence. Gil-White gives a telling historical example to illustrate the point that is so germane to our main topic that I will quote it at length.

> The Weimar Jews were quite assimilated to German society in speech, custom, and dress, had fought as Germans in World War I, and, without relinquishing a Jewish identity, often considered themselves genuine Germans. But in the ensuing anti-Semitic rampage not merely those who preserved Jewish ascriptions and traditions for themselves but even those with a small fraction of Jewish ancestry (sometimes as little as one-eighth) were slated for persecution. Nazi anti-Semitism openly essentialized its victims, attributing to them a corrupt nature. Not all were convinced by this ideology to the point of justifying the persecution of Jews, but the question remains: Why was it so plausible to Nazi converts that even a little bit of Jewish "blood"—unknown perhaps even to their bearers and against all the powers of German enculturation—would pass on this supposedly corrupt nature? Perhaps because we intuitively process ethnic groups as though they were "species," reasoning implicitly that the corresponding nature is passed down reproductively, and, hence, in the "blood."[6]

This is an extremely important point. We naturally tend to think of the members of human communities as sharing a common essence—often thought to be something in the "blood." Of course, this is a false but powerful idea that is crucial for understanding human xenophobia, nationalism, racism, and related phenomena. Imagined communities are held together by such mythic commonalties.

With this in mind, the differences between the human and chimpanzee patterns of aggression no longer seem so absolute. Chimpanzees are prone to attack individuals outside their *physical* communities, but usually avoid lethal aggression against insiders. Humans take a similar stance toward those inside and outside their *imagined* communities, but members of a single physical community often belong to diverse imagined communities, which makes them potential targets for violence. The Hutu and Tutsi of Rwanda are a case in point; these people lived side by side, but in 1994 almost one million Tutsi were literally butchered—chopped to bits by machete-wielding Hutu—because of their essence: *they were Tutsi*. What divides can also unite. Our ability to form alliances between groups is an upshot of the same principle that tears some groups apart. *Communities unite by treating one another as comembers of a larger imagined community that shares a common essence*: "Christian or Jew, black, white, or Hispanic, Democrat or Republican, we are all Americans."

If I am right, human warfare must have followed a chimpanzee-like pattern until our ancestors reached the level of cognitive sophistication that allowed them to form concepts and to attribute essences to things, and then, for the first time, human conflict acquired a potent ideological dimension. New notions of tribal, ethnic, and religious identity emerged, groups began to forge alliances with one another, and ideas about an immortal human essence made men willing to die in battle. True warfare began. "This," observes Gwynne Dyer, "was violence pushed to the limit with an almost Clausewitzian* determination, and it was there right at the start of civilization."

* Karl von Clausewitz (1780–1831) was a Prussian general and military theorist famous for his conception of "total war."

> Maybe the cumulative losses in the hunter-gatherer and tribal style of warfare were still proportionally greater over a generation then those suffered by Mesopotamian city-states in war . . . but the intensity of the fighting, the willingness of large numbers of men to stand their ground and fight, even given the high probability that they would die there in the next five minutes, had no precedent in the human, primate or even mammalian past. . . . It was culture that made their sacrifice possible.[7]

However, there was a monkey wrench jammed in the well-oiled machinery of slaughter. Before they crossed the military horizon, our ancestors had no scruples about killing strangers: they played the game of life by chimpanzee rules. But once they mastered conceptual thought and began to attribute essences to things, it gradually began to dawn on them that, despite superficial differences in appearance, language, and dress, all human beings are members of a single kind. This realization activated the deep inhibitions against killing members of the group inherited from our prehuman lineage. Paradoxically, the same cognitive mechanism that made true war possible also made it difficult to pursue. Ancestral human beings were caught in a bind. Both natural and sexual selection had bred ferocity into them, but these aggressive urges, which had served us so well in the past, were now thwarted by an equally profound aversion to killing.

To understand how our species escaped from this paralyzing bind, we must delve once again into the workings of the human mind.

MIND READING

Imagine that you are a traveler to a distant planet—a planet that supports life but (in the words of the *Star Trek* cliché) "not as we know it." The animals on Planet X look nothing like earthly animals. They do not have faces, legs, or tails; fur, feathers, skin, or scales. How do you think that you would recognize them as animals? The odds are that you would attend to the way that they *move*. The human brain appears to be hard-wired to distinguish between animals and nonanimals, and it does this largely by tracking their locomotion. Animals are self-propelling—they can start, stop, and change direction as they choose. A mouse scurrying around the top of a pool table behaves in a totally different manner than a cue ball. The mouse starts and stops, accelerates, decelerates, and changes direction in apparent pursuit of a goal. The ball does nothing unless some force is brought to bear on it, say, by a cue stick, and when it moves, it moves in a way that is linear and utterly predictable. (Imagine how hard it would be to play pool if this were not the case!) The movement of the ball is completely determined by forces acting on it from the outside, whereas the mouse seems to be powered by an internal motor, by a "spirit" (in Latin, *anima,* hence the word "animal"). When we observe a thing that moves more like a mouse than like a ball, a neural switch gets flipped inside of our brains, which causes us to see the thing as an *agent,* an entity that moves through the world under the influence of its inner states.[8]

When we perceive something as an agent, we automatically adopt what Daniel C. Dennett calls the *intentional stance* toward it.[9] To take an intentional stance towards something is to credit it with mental states which we use to explain and predict its behavior. There is a sense in which the mouse's movements are unpredictable. However, at a more abstract level, its actions (or, if you prefer, its behavior—a word which has a peculiar aura of scientific legitimacy denied to words like "action" in social science circles) are remarkably predictable. Imagine that you know that the mouse has been deprived of food, and you notice that there is a piece of bread lying on the table. It's child's play to predict that the mouse will approach the bread and eat some of it. This kind of prediction does not require fancy scientific theories, mathematical models, or measurements. All that's required is common sense: we assume that the mouse is *hungry,* that it *desires* food, that it *believes* the bread is food and therefore that it will *eat* some of the bread. In ordinary life we almost never spell out such assumptions so explicitly. However, unless we had something like them in mind, it would be impossible to make sense of animal (and therefore human) behavior.

Do mice *really* have beliefs and desires? Who knows? In fact, it's not clear what "really" having beliefs and desires, as opposed to "sort of" having them, *really* means. Describing the mouse's behavior using a grid of intentional states is at least as useful as describing the location of a town in terms of latitude and longitude. Do latitude and longitude *really* exist? Yes and no. Although the equator is not "real," you can still be north or south of it. The equator is not like the Mississippi River. It is not a feature of the earth. Rather, it is part of a conceptual framework that we use to describe the earth. The same considerations apply to the mental states that we attribute to other animals. It is not possible to know what, if anything, goes on in a mouse's subjective experience. But we do know that attributing beliefs and desires to it enables us to describe and predict its behavior, whereas taking the very same stance toward a billiard ball gets us nowhere fast.[10]

Having the ability to take the intentional stance strengthened our ancestors' survival prospects. As we have seen, early humans were relatively slow, not particularly strong and had no protective hide and no natural weapons like horns or large canine teeth to protect them from predators. But they had one great asset: their brains. They relied on their wits to keep one step ahead of their natural enemies. Being able to infer what a leopard has in mind and to use this inference to anticipate its next move is a very useful talent. It also pays big dividends when hunting for game (you stand little chance of catching the rabbit unless you are a good-enough rabbit psychologist to be able to figure out how it is likely to behave as you approach it). As cognitive archeologist Steven Mithen observes: "Anthropomorphizing animals by attributing to them human personalities and characters provides as effective a predictor for their behavior as viewing them with all the understanding of ecological knowledge possessed by Western scientists."[11]

The important point in all of this is that we automatically attribute mental states to other beings as soon as the neural switch gets flipped.

Of course, some scientists have flatly denied that nonhuman animals have mental states. Descartes, for example, described nonhuman animals as glorified clockwork toys, and as recently as the twentieth century, behaviorists claimed that there was nothing more to animal behavior than bundles of mindless reflexes (to be fair, they also described human beings as consisting of nothing more than bundles of reflexes). Were people like Descartes and the eminent behaviorist B. F. Skinner immune from the automatic response described above? I doubt it. It is more likely that these people's theoretical beliefs have overlaid their natural, human attitude. Of course, anyone can *claim* to have any belief they want. You can deny the omnipresence of gravity, but I bet it won't lead to your jumping off the balcony. But our genuine beliefs—the beliefs that we live by—are bound to show up in our behavior, and I doubt very much that denials of animal mentality ever get farther than the study door. When Descartes let down his hair and knocked back a few glasses of Beaujolais he probably said, "Good boy!" to a friendly dog just as readily as the next man, and even the most hard-boiled behaviorist returning home after a day in the lab just can't help recognizing that Fido is yelping excitedly and wagging his tail because he is *glad* to see his master.[12]

Psychologists call this talent "mind reading"—which sounds a bit spooky. There is nothing supernatural about mind reading, in the psychologists' sense of the term. It refers to our refined ability to accurately attribute mental states others. Most of our mind reading is directed toward other people rather than toward nonhuman species. Each of us goes through life interpreting other people's behavior, attributing beliefs and desires to them and using this to anticipate and understand their behavior. Mind reading happens spontaneously—we don't have to concentrate or make any kind of mental effort to do it. When Sasha bumps into her old friend Prerna and Prerna smiles broadly and extends her arms, Sasha immediately knows that Prerna is overjoyed to see her. Prerna 's feelings seem to be written all over her face. Sasha can see this because she possesses a module in her brain that is specialized for interpreting other people's mental states, and her seemingly instinctive insights into her friend's mind are actually mediated by complex cognitive processes churning away below the threshold of awareness. The same kinds of processes occur in your mind when you interact with other people. Imagine what your life would be like if you couldn't do this. You would be unable to understand the meaning of human behavior and would perceive people as hunks of flesh moving mindlessly through space. Here is University of California psychologist Alison Gopnik's vivid description of what this might be like:

> This is what it's like to sit round the dinner table. At the top of my field of vision is a blurry edge of nose, in front are waving hands. . . . Around me bags of skin are draped over chairs, and stuffed into pieces of cloth, they shift and protrude in unexpected ways . . . two dark spots near the top of them swivel restlessly back and forth. A hole beneath the spots fills with food and from it comes a stream of noises. Imagine that the noisy skin-bags suddenly moved toward you, and their noises grew loud, and you had no idea why, no way of explaining them or predicting what they would do next.[13]

Gopnik's portrait is not just a psychologists' fantasy. It is her imaginative leap into the subjective world of an autistic person. According to the prevailing view, individuals suffering from autism do not possess functioning mind-reading modules: they live in a world in which nothing has a mind.

How about nonliving things? Can they have minds? The notion of machines with minds has been standard science fiction fare for decades. Thanks to the rapid pace of technological development, idea of an intelligent machine—a computer endowed with "artificial" intelligence—no longer seems far-fetched. But what exactly *is* machine intelligence, and how can we know when we have met it? The brilliant mathematician Alan Turing published the classic discussion of this issue in 1950, at a time when the computational wizardry to which we have grown accustomed today was barely imaginable. Turing invited his readers to picture a human interrogator sitting at a computer terminal through which she interacts with a computer locked away in another room. The interrogator has no idea whether the entity with which she is conversing is a computer or another human being, and her task is to find out. The computer has been programmed to simulate the human conversational behavior, so *its* task is to deceive the interrogator. A machine that can pull this off is said to have passed the Turing test and should, according to Turing, be considered an intelligent being.[14] Turing's thesis has been discussed and debated in the literature for over fifty years, and a variety of objections have been leveled against it. Most of these boil down to the idea that any computer that passed the Turing test would only be giving a good imitation of intelligence rather than displaying the genuine article. This objection misses the point by a wide margin. A system's behavior is the *only* basis for judging whether or not it is intelligent. It follows that, if a computer's behavior is *indistinguishable* from the behavior of a human being, then if we deny that the computer is intelligent, we must—on pain of inconsistency—also deny that human beings are intelligent. What's sauce for the goose is sauce for the gander, and the issue of real versus simulated intelligence is a spurious one.[15]

Nowadays, most educated people are probably comfortable with the idea that computers can, in some sense, think. Far fewer are ready to take on board the idea that they can have minds. What's the difference? Well, concluding that a computer can think is a *theoretical* judgment, like the behaviorist's judgment that animals don't have mental states. It comes from the head, not from the guts. When we reason that a computer can think, this is very different from the *feeling* that it has a mind. When confronted with a computer, even a clever chess-playing one, we are left cold. The machine's feats are impressive, but there seems to be something vital missing: computers just don't seem to be the kind of things that can have minds.

Let's extend the Turing's scenario to explore this idea more fully. Imagine that *you* are administering the Turing test and a computer program manages to convince you by means of its witty repartee that it is really a person. After the test is finished, a laboratory assistant escorts you to the other room to meet your conversational partner. Entering the room, all that you see is a metal box with some flashing lights sitting inertly on top of a desk. How would you react? I bet that you would be surprised, and after getting over the initial shock, you would reluctantly concede that the computer must be intelligent even though there is something unnatural about this idea. Now, imagine that instead of seeing a PC on the desk you are greeted at the door by a cute little robot like 3CPO. Would it be any easier to think of this machine as an intelligent being? I bet it would, because in this realm looks matter . . . a lot. An immobile computer on a desk would not activate your mind reading module, and you would not get that *visceral* sense of the presence of another mind. Mother Nature designed our mind-reading equipment to respond to specific visual cues that were reliable indicators of intelligence in the prehistoric environment in which we evolved—an environment in which smart computers did not exist, and animals were the only intelligent systems around.

Treating a creature as an agent presents no obstacle to killing it: we cheerfully scoff down live oysters, drop lobsters into boiling water, gut gasping fish, and commit a range of similar atrocities against our nonhuman relatives. This shows that merely having a mind is not enough to grant a creature citizenship in our moral universe (if the Sixth Commandment applied to chickens, Colonel Saunders would be the worst mass murderer in history). This is not just a twenty-first-century affectation: it goes back to ancient times. In the fourth century B.C. Aristotle wrote that nonhuman animals should be used as "tools," and the Roman emperor Marcus Aurelius advised his readers to "use animals and other things and objects with a liberal spirit, but towards human beings as they have reason, behave in a social spirit."[16] This instrumentalist approach to nonhuman creatures was evident in the Roman arena, where wild animals were slaughtered to entertain massive crowds. Torturing animals was a also popular entertainment through the Middle Ages and even during the Age of Enlightenment.

> At the courts of royalty, lords and ladies enjoyed watching assorted large animals—lions, bears, horses, wolves,

> bison—goaded into fighting and killing each other in roman-style arenas. In the establishments of princelings too poor to slaughter lions and bison for entertainment, courtiers gathered in rings to toss small animals in nets until they died from accumulated fractures, concussions and shock. . . . Commoners amused themselves will bull-baiting and bear-baiting, bull, dog, and cock fights, throwing at cocks (hurling sticks to maim and kill tethered cocks), and riding at geese (riding at a gallop to pull the head off a greased goose hanging in a tree). Meat animals were tortured to death in various ways to make their flesh more tender and savory. Cats were routinely set on fire, hung up in bags and smashed like piñatas, or killed in dozens of other ingeniously painful ways for recreation.[17]

The practice of torturing animals for sport persists to this day in activities like hunting, fishing, bullfighting, and cockfighting. We butcher them by the millions to satisfy our desire for meat, leather, and other useful products, but the prospect of killing *human* animals normally leaves us aghast. When I used the examples of oysters, lobsters, and fish to show that taking the intentional stance toward an animal is compatible with killing it, I intentionally confined myself to some pretty low-level agents. What happens when we shift the focus to the mammals? Most of us with no moral qualms about boiling a lobster would have difficulty slaughtering a lamb. Some people object to slaughtering lambs as a matter of principle, but most do not. They are the sort of people who do not hesitate to dig into shish kebabs and other ovine delicacies with gusto. These people (and I count myself among their number) don't object to the killing per se, they just don't want to be the one who has to do it.

PEOPLEMAKING

Why do you suppose that is it emotionally easier to kill a lobster than it is to kill a lamb? David Hume suggested the answer, when he noted that people are biased in favor of those individuals that resemble them. Of course, Hume had human beings in mind, but the same principle seems to apply across species. Generally speaking, *we have difficulty killing animals to the extent that they resemble us.* Lambs are mammals, and are a lot more like us than lobsters are. Lobsters are hard and cold, they live under water, have lots of legs, and don't really have faces to speak of. Lambs are warm and furry, have forelimbs and hindlimbs, have voices, and are more or less the size of a human child. Most importantly, they have *faces.* Faces rank very high on our list of visual priorities. In social interactions, we fix our gaze on people's faces. We read their emotions from their facial expressions (smiles and scowls, eyes that light up with joy, widen with surprise or darken with anger). Even tiny babies gaze more intently at human faces than they do at other objects. "The recognition of the human face," writes art historian Ernst Gombrich, ". . . is based on some kind of inborn disposition. . . . Whenever something remotely facelike enters our field of vision, we are alerted and respond." Face recognition even has a special niche in the neuroanatomy of the brain (a little strip of tissue at the bottom of the brain called the fusiform gyrus).[18]

Narrowing the focus still further, the mind-reading module is especially sensitive to the sight of humanlike *eyes.* Eyes have always been special for primates. Unlike other mammals, primates have eyes at the front of their heads, rather than the sides, which greatly enhances their depth perception. This came in handy to our distant, tree-living ancestors, but to gain the benefit they had to pay a price. As their eyes rotated to the front of their heads, primates gained focus but lost the panoramic view of the environment available to other mammals, and this made it much more difficult to for a lone individual to spot predators. This favored a communal lifestyle, in which many pairs of eyes looking in many directions at once can compensate for the reduced scope of each individual pair. Many other animals track the gaze of other individuals, but in our species, vision has become so important for interpersonal communication that the brain has evolved a special subsystem, wired to the social brain, that allows us to make subtle inferences about the mental states of others just by looking into their eyes.[19] Looking into a pair of human eyes is a very different experience from looking into a fish's eyes. Human eyes—or rather human-looking eyes—give the impression that there is someone at home. The old adage that the eyes are the mirror of the soul goes back to at least the second century B.C., when Cicero wrote "Ut imago est animi voltus sic indices oculi" (The face is a picture of the mind as the eyes are its interpreter). If the eyes are mirrors of the soul, then we should be able to determine whether or not a creature has a soul by looking into its eyes. We've already seen that the "soul" is probably a folk term for the human essence: a creature with a soul, then, is a creature that is conscious, self-aware, a subject, a being with an inner life. It is one of us: a person.

You can perceive a person's size, shape, complexion, eye and hair color, and, with suitable instruments, their weight, body mass index, and so forth, but you cannot perceive their consciousness. Consciousness is something that we *attribute* to others on the basis of certain cues. We perceive others as conscious entities—as subjects—to the extent that they have human-looking faces and, especially, human-looking eyes. If you doubt this, consider the following modification of our Turing test vignette. Having been persuaded that you were chatting with a real, flesh-and-blood human being, the lab assistant takes you behind the scenes to introduce you to your interlocutor. She opens the door and this time, instead of a little 3CPO-style robot, you are greeted by a robot topped with a simulacrum of a human head, complete with glossy hair, moist eyes, and warm skin. I believe that you would not only easily grant that the robot has a mind, but you could not help perceiving it as a conscious being. The sight of a human face causes us to see the bearer of the face as a *subject*. We even attribute consciousness to two-dimensional images on the television screen. The sheer automaticity of this effect suggests that it is hard-wired into us, the work of a dedicated mental module. I call it the people-making module. Couldn't the peoplemaking effect be a response to the general human form rather than a specific effect of viewing the face? I doubt it. If your conversation partner in the Turing test turned out to be a centaur robot (or, indeed, a genuine centaur), with a horselike body but a human-looking head I suspect that it would produce much the same effect. But a Minotaur robot, a robot with a human-looking body and a bull-like head, would not. This is why severe facial disfigurement is such a dreadful disability. Extremely misshapen faces fail to light up our people-making modules, and we tend, tragically, to treat these individuals as less than human. In one of the most moving scenes in David Lynch's film *The Elephant Man* the deformed Joseph Merrick cries out, "I am not an animal. I am a human being." Although this makes no *biological* sense, it makes tremendous psychological sense. Merrick was trying to convey something about his *moral* status. In saying "I am a human being," he was saying something like "Despite my appearance, I have a human essence. I am a conscious being like you, deserving of respect and compassion." This is precisely the joint at which the folk distinction between "human" and "animal" is sliced. Only beings that activate our peoplemaking module are felt to be people. Merrick had to *assert* his human status because his deformed face did not cause others to spontaneously recognize it. Being human in this sense is not absolute: it comes in degrees. A lamb is more human than a lobster, and a chimpanzee is more human than a lamb.

THE EYES OF THE ENEMY

The sight of a human face switches on the peoplemaking module, arouses compassion, and restrains aggression, and this poses a problem for the would-be warrior because it interferes with the business of killing. Of course, all soldiers *know* that they are fighting against other human beings. But this is different from the instinctive response prompted by seeing an enemy soldier's face. This is why the sensible command "Don't fire till you see the whites of their eyes" is hard to put into practice. When you see the whites of the enemy's eyes, you also see that that your enemies are people.[20] It is far easier to kill a faceless person. Dave Grossman explains that this is why "executions are traditionally conducted with a bullet in the back of the head, and why individuals being executed by hanging or firing squad are blindfolded or hooded. And we know from Miron and Goldstein's . . . research that the risk of death for a kidnap victim is much greater if the victim is hooded."[21]

One solution to this problem is to recruit only psychopaths into the armed forces, people who are unable to feel sympathy for others and can look another person in the face and then blow his head off. As one World War I military psychologist bluntly put it, the ideal soldier is "more or less a natural butcher."[22] But psychopaths are a scarce commodity and make up only a very small percentage of the armed forces of any nation. They are also dangerous to their own comrades, because they are immune to the feelings of loyalty and affection that unite and motivate a fighting unit.

Another approach is to kill at a distance, because a soldier who is far enough away cannot see his victim's face and can avoid recognizing the humanity of his targets.[23] This approach was pioneered by the fourteenth-century English, whose archers, equipped with longbows, mowed down French troops during the Hundred Years War, and it was enhanced by the invention of firearms that could effectively kill at a distance. As one veteran of the Second Gulf War put it, "When you're in a tank battalion, you're not shooting at vehicles with four people in it. You're shooting at tanks. We kill tanks, not people."[24] Long-range killing reached its apogee when we developed the capability of dropping bombs from airplanes flying high above their targets. It is interesting that Lieutent Colonel Paul Tibbets, the pilot of the plane that dropped the atomic bomb on Hiroshima that killed approximately 92,000 people, remarked afterward that he felt no remorse. In fact, in a 2002 interview the ninety-year-old Tibbets told journalist Studs Terkel that he would not hesitate to use nuclear weapons against Al Qaeda.

> I'd wipe 'em out. You're gonna kill innocent people at the same time, but we've never fought a damn war anywhere in the world where they didn't kill innocent people. If the newspapers would just cut out the shit: "You've killed so many civilians." That's their tough luck for being there.[25]

The crucial role of distance in releasing inhibitions against aggression also explains why politicians and the general populace are typically much more keen on slaughter than soldiers on the front lines are (today's conservative radio hosts bay for blood from the comfort of their air-conditioned studios, while the grunt on the ground in Iraq or Afghanistan is often less zealous). Distance also explains how it is possible for generals to condemn thousands of men, women, and children to death with a single order without suffering psychological harm, while a private who kills just one man in close combat may be haunted by the experience for the remainder of his life. Do you remember the trolley problem briefly mentioned in chapter 7? Most people were willing to pull the lever to divert the train, even though this results in the death of a pedestrian, but they were far less willing to push a fat man in front of the trolley even though it would result in the same net loss of life. The prospect of shoving the man onto the track is, I believe, just too *intimate* an act of killing for most of us to contemplate without distress. Indirect, impersonal killing has little if any effect on the soldier's psyche. It is when the slaughter is up close and personal, and the humanity of the victim cannot readily be denied, that our profound aversion to taking a life swings into action, *and must be overcome if the soldier is to perform his lethal duty*.

At first glance, snipers seem to be an exception to the rule that distance makes killing easier. But this is more apparent than real. Although they kill from a distance, the telescopic lens on a sniper's rifle brings him *visually* close to his victim, so, in effect, a sniper is someone who kills up close without exposing himself to the risks of hand-to-hand combat. Because of this, snipers have often been detested by their comrades in arms, and they are also especially vulnerable to feelings of guilt. As one World War II sniper put it, "The game was dirty . . . the cool calculating murder of defenseless men was diabolical."[26]

The horror of combat is captured in the *faces* of dead enemies, and references to them are frequently found in accounts of the experience of war. But the most traumatic form of killing involves looking a person in the eyes while taking his life. It is an experience that often haunts men for the rest of their lives. The experience of this most intimate form of lethal violence is illustrated by a remark made to French journalist Jean Hatzfeld by one of the Hutu killers interviewed in his disturbing book *Machete Season*: "The eyes of someone you kill are immortal if they face you at the fatal instant. . . . They shake you more than the streams of blood and the death rattles, even in a great turmoil of dying."[27] Lieutenant Colonel Dave Grossman makes the point with particular clarity in his book *On Killing:*

> Looking another human being in the eye, making an independent decision to kill him, and watching as he dies due to your action combine to form the single most basic, important, and potentially traumatic occurrence of war. If we understand this, then we understand the magnitude of the horror of killing in combat.[28]

We have seen that sheer physical distance makes it much easier to kill another person, because the specific stimuli that switch on the people-making module are not present. Subjectively speaking, people killed at a distance are not experienced as real people. They are "targets," or dots on the landscape. It is partially because of this that the use of aerial bombardment, artillery, chemical weapons and other long-range devices have probably had a disproportionately deadly impact on warfare. But even in modern warfare, not all fighting can be done from a distance. As we have seen, some soldiers deal with this by creating a *psychological* distance between themselves and the enemy. They become dissociated and experience the enemy combatants more like two-dimensional movie characters than sentient beings. Both forms are distancing are, in the final analysis, forms of self-deception. They are ways of creating and sustaining the illusion that one is not taking human life. Physical distancing is a technologically mediated form of self-deception. A person who drops a bomb from an airplane can pretend that he is not really killing anyone. Dissociation is a form of neuropsychologically mediated distancing, but the end result of both is much the same. We will see in chapter 10 that there are other, very dangerous forms of psychological distancing that also rely on the self-deception to circumvent an awareness of the enemy's humanity.

10

PREDATORS, PREY, AND PARASITES

> He who fights with monsters might take care lest he thereby become a monster. And if you gaze for long into an abyss, the abyss gazes also into you.
>
> —FRIEDRICH NIETZSCHE, *BEYOND GOOD AND EVIL*

PSYCHOPATHIC THUGS DO NOT FIGHT wars. Ordinary people do, and although war is all about killing, we do not like to think of our ordinary people—"our boys"—as professional killers. As Bertrand Russell once observed, we are quick to say that they give their lives for their country but not that they *take* lives for their country. To do so would be to upset the moral order of things. However, we have no such reservations when it comes to the enemy. There are no heroes on the other side, no brave young patriots making the ultimate sacrifice for their country. The enemy is ruthless and diabolical; he is a terrifying, cold-blooded killer. "When our own nation is at war with any other," observed David Hume in the year 1740, "we detest them under the character of cruel, perfidious, unjust and violent: but always esteem ourselves and allies equitable, moderate and merciful. If the general of our enemies be successful, 'tis with difficulty we allow him the figure and character of a man."[1]

Notice that Hume remarked of the enemy that it is only "with difficulty that we allow him the figure and character of a man." This is perhaps the first published observation of *dehumanization* in war.

Nowadays it is widely accepted that we tend to picture our enemies as less than human—so widely accepted, in fact, that it has become a cliché. Like all clichés, we seldom if ever pause to consider it seriously. I think that the notion of dehumanization in war contains a profound and extraordinarily important insight into human nature. In light of the points raised in the previous chapter, it is easy to see how dehumanizing the enemy would provide an elegant solution to the problem posed by our innate aversion to taking human life. Perceiving the enemy as nonhuman would liberate us from inhibitions against killing them. A perceptual shift of this nature would enable human beings to take the lives of others as casually as they would swat a mosquito, poison a rat, or impale a writhing worm on a fishhook.

Social scientists have taken up Hume's insight, and extended it in small but significant ways. As early as 1911, the American sociologist and political economist William Graham Sumner noted that "savage tribes" often refer to themselves by names that mean "men" or "the only men" and that this implied that outsiders are not truly human. Seven years later, during World War I, a British psychologist named John T. MacCurdy linked dehumanization with primitive warfare more explicitly, when he claimed that primitive people perceive their enemies as nonhuman animals. The idea cropped up again in 1964, in a book by the psychoanalyst Erik H. Erikson, who called it "pseudospeciation." This term was adopted by the Austrian biologist Irenäus Eibl-Eibesfeldt a few years later in an influential book on *The Biology of Peace and War* in which he made the point that a belief that outsiders are not really human liberates us from deep inhibitions against killing our own kind. This was a very important insight, but there was still something lacking in the analysis. How is it possible for normal, intelligent human beings to believe that their enemies are not really human? This perplexing question was not resolved until 2001 when Francisco Gil-White advanced the argument that the human brain has a module for thinking about human groups. This module conceives of ethnic groups as sharing an invisible essence that is "hidden" inside of a person. At the gut level, we all tend to categorize human groups in this manner, because it is an unconscious, automatic tendency etched into our brains by evolution. How did this come about? Gil-White argued that it is an extension of a system that evolved to help our ancestors categorize *kinds* of living things. As a result, we treat members of alien ethnic groups as though they were members of a different species. We may notice that members of another group look and dress and speak exactly as we do, and yet we may feel that there is something that is different about them. Something invisible, impalpable, and hard to pin down. Something at their core or in their "blood" that distinguishes them from us. This principle is well illustrated by a remark made to my mother decades ago when she was one of the very few Jews living in a small Southern town. She was told, "A Jew's just a nigger turned inside out."

So, how does all of this help us grapple with the psychology of war?

There are various forms and degrees of dehumanization. United States soldiers in Iraq sometimes call Iraqis "hadjis," "ragheads," or "camel jockeys." These are derogatory terms that create psychological distance. Clearly, it is easier to do violence to a raghead than it is to harm a full-fledged human being. But these are not especially nasty epithets. They may contribute to an attitude of callousness, but they do not inspire hate, fear, or repugnance. Not all instances of dehumanization are so moderate. Soldiers sometimes imagine their enemies as dangerous, subhuman beasts. As Vietnam War veteran Bob McGowan explained to CNN, "They're subhuman. They're animals. They're going to rape our women and kill our children. . . . Kill them."[2] This is a step beyond dehumanizing the enemy: it is *demonizing* them.

When we demonize others, we perceive them as having a *dangerous nonhuman essence*. They are, so to speak, wolves in sheep's clothing. The idea of a human-looking entity with an evil nonhuman essence is a staple of folklore, horror films, and religious mythology. In *The Exorcist*, for example, a child's body is occupied and controlled by a malevolent demon. Dracula, too, looks like a person but is actually one of the "undead." And when Jesus drove the Gadarene swine over the precipice to their death, they had the outward form of pigs, but the hidden essence of a multitude of demons. The concept of an evil being concealed within a human body comes quite naturally to us. One of the most instructive examples of this kind of essentialist thinking is the medieval notion of the changeling. In her wonderful book *Mother Nature*, anthropologist Sarah Blaffer Hrdy describes how medieval parents managed to convince themselves "that sickly babies were imposters left by goblins in place of healthy ones."

> The infant left behind became an *enfant changé* in France, a *Wechselbag* in Germany, in England a "fairy child" or changeling. In the best known versions from northern Europe, changelings were left overnight in the forest. If the fairies refused to take it back, the changeling would die during the night—but since it was not human, no infanticide could have occurred. If by a miracle the exposed baby survived, it meant that the original healthy human child must have been returned.[3]

The changeling was a being that looked like a baby, sounded like a baby, smelled like a baby, and behaved like a baby but, although it was physically indistinguishable from a real baby, *it was inwardly demonic and therefore could be killed without guilt or remorse.*

Hrdy tells us that in most cultures, a mother is allowed to kill her baby shortly after it is born if she deems this appropriate. However, this must always be done prior to the naming ceremony, which is when the baby is thought to acquire a soul. Because medieval Europeans were Christians, they believed that babies are born with souls. Consequently, they had to concoct the elaborate self-deceptive fable of the fairy child, which, by denying the baby's humanity, licensed their killing it. The problem faced by medieval parents of unwanted children is very much like the problem confronting men at war. The soldier, too, has a reason to kill. But his awareness that his enemies are human beings makes it difficult for him to do his job. As we will shortly see, many soldiers also adopt a similar solution. They deceive themselves about the moral status of their enemies, slipping into the view that although the enemy *looks* human, he or she is not *really* human.

There is a crucial difference between the case of the medieval mother and the warrior on the battlefield. Medieval mothers did not actively kill their infants; they abandoned them to die. But the soldier must deliberately *kill* the enemy. These may seem to boil down to the same thing—surely, you may think, leaving a baby to the wolves is tantamount to killing it. This is true, but there is an important psychological distinction between the two acts. Remember the two trolley problem scenarios described in chapter 7? An act of killing that requires you to get your hands dirty—in this case, to push the fat man onto the tracks—has a very different emotional valence than pulling a lever. Although dehumanizing his enemy goes a long way toward freeing him from the oppressive burden of conscience, it helps even more if a soldier perceives his enemies as the kind of nonhuman beings that must be met with violence. Killing other people is easiest if there is something about them that makes you *want* to kill them—something that arouses deep aggressive passions.

INVOKING THE GHOSTS OF PREDATORS PAST

Let's begin by looking at how enemies are represented in wartime propaganda. I have chosen five examples that I think are especially revealing.[4]

- A Union poster from the American Civil War shows a heroic, club-wielding General Scott of the Union army. He is poised to bludgeon a gigantic, nine-headed serpent. Seven of the monster's heads are those of leaders of the Confederacy.
- An American cartoon from the Spanish-American War represents Cuba as a huge, sinister ape-man, complete with protruding fangs, holding a bloody knife and hulking

over the grave of U.S. servicemen killed in the battleship *Maine*, which blew up in Havana Harbor in 1898.

- A Soviet poster from the cold war shows two elderly women and a man holding a child in the foreground. A huge atomic bomb equipped with gaping jaws and huge canine teeth looms in the background. The picture is captioned "They are threatening us."
- A Taiwanese cartoon depicts a hapless man in a wooden boat about to be devoured by an enormous shark. The boat is labeled "Taiwan" and the shark is labeled "China."
- A Soviet poster from the 1950s shows a wolf dressed in a suit and tie removing a mask from its grotesque, snarling face. The mask bears the countenance of United States Secretary of State Dean Acheson.

In each of these examples a human enemy is depicted as a nonhuman predator. Many, many other examples can be found in political poster art, which often represent national enemies as birds of prey, wolves, bears, tigers, and sharks, and sometimes huge serpents and gargantuan spiders.[5] The obvious message is that our enemies are both nonhuman and dangerous. They are animals which, if not killed, will devour us.

As the cannibalistic barber in Steven Sondheim's musical *Sweeney Todd, the Demon Barber of Fleet Street* reminds us, "The history of the world, my sweet, is who gets eaten and who gets to eat."[6] This insight is lost on most of the denizens of the developed world who sit comfortably at the apex of the food chain. We do the eating. We slaughter millions of animals every day to satisfy our appetite for meat, but we are unlikely ever to feel an animal's jaws close on our flesh. We are so accustomed to being safe from nonhuman predators that when a cougar mauls a California jogger or a shark bites off the limb of a surfer, the story makes the national news. But don't be fooled. Even today, predatory danger is a real problem in some parts of the third world. Donna Hart, an anthropologist who studies predation on humans, recently wrote:

> Although we are not likely to see these facts in American newspaper headlines, each year 3,000 people in sub-Saharan Africa are eaten by crocodiles, and 1,500 Tibetans are killed by bears about the size of grizzlies. In one Indian state between 1988 and 1998, over 200 people were attacked by leopards; 612 people were killed by tigers in the Sundarbans delta of India and Bangladesh between 1975 and 1985. The carnivore zoologist Hans Kruuk, of the University of Aberdeen, studied death records in Eastern Europe and concluded that wolf predation on humans is still a fact of life in the region, as it was until the 19th century in Western European countries like France and Holland.[7]

Although most of us are safe from predators, every year we spend millions of dollars to sit in darkened rooms and observe other people being stalked, chased, and eaten. High-grossing movies like *Jaws, Alien, Predator, Jurassic Park,* and *War of the Worlds* arouse primal emotions within us. We respond to these films because they resonate with ancient fears of being hunted and eaten, fears that were installed in our minds during the millions of years during which our ancestors lived in terror of the deadly, flesh-eating monsters with which they shared their world. "Our fear of monsters in the night," wrote environmental philosopher Paul Shepherd, "probably had its origins far back in the evolution of our primate ancestors, whose tribes were pruned by horrors whose shadows continue to elicit our monkey screams in dark theaters. As surely as we hear the blood in our ears, the echoes of a million midnight shrieks of monkeys, whose last sight of the world was the eyes of a panther, have their traces in our nervous system." Our ancestors had to constantly be on guard to protect themselves from crocodiles, hyenas, big cats, bears, and wolves. "Death by predation must have seemed ordinary," remarks science writer David Quammen. "No one had escaped the awareness of being meat."[8]

Long ago, before our ancestors learned to make weapons and hunt effectively in groups, they were dependent on predators for meat. Like jackals and vultures, they lived—in part—as scavengers, dining on big cats' leftovers—when they could get them. These, of course, were the very same predators that stalked and ate human beings. As a result, according to the intriguing thesis set out by Barbara Ehrenreich, prehistoric people developed a curiously ambivalent relationship with these creatures. On the one hand, they were terrified of them, but on the other, they venerated them.[9]

Evolution shaped the primate psyche to fear large, aggressive animals that might attack and devour them. Over millions of years, this constant danger exerted a huge selection pressure and, as a result, our ancestors evolved a mental module specialized for dealing with predators. This module is switched on when we watch films like Ridley Scott's *Alien* ("In space no one can hear you scream") or when we hear the sound of footsteps behind us while walking on a deserted street late at night. But for our prehistoric ancestors (and for contemporary people living in areas of the world where predation is still a problem) the predator-detection module was and is vitally important. It made a difference between life and death.

At this point, it makes sense to briefly revisit and expand upon some points about human evolution that were raised earlier in this book. As I described in chapter 1, at some point in prehistory—we do not know exactly when—our ancestors were able to turn the tables on the fearsome creatures that hunted them. They learned to craft weapons and hunt in well-organized bands and became more than a match for their animal foes. When our forebears reached this stage of social organization and technological prowess, they were faced with a new and more formidable danger: the danger posed by one another. Armed and dangerous, rival human communities began to clash in ways that were more lethal than the territorial skirmishes of chimpanzees. Imagine chimpanzees armed with bows and arrows, spears, stone axes, and . . . language! Chimpanzees who were able to collectively plan aggressive strategies, ambushes, and massacres and possessed the cognitive and communicative wherewithal to carry them out. The early warriors must have taken the skills that they evolved in their long struggle against the predations of leopards and saber-toothed tigers and redeployed them against human animals.

Ancestral experiences can leave deep imprints on an animal's body and mind. Evidence suggests that the remarkable sharpening of primate vision and the rotation of primate eyes from the sides to the front of the skull were adaptations for detecting the large snakes that fed on them. Long ago, pronghorn antelope were chased by cheetahs across the plains of the American west. The cheetahs are no more, but the antelopes' speed remains. In the evocative words of biologist John A Byers, pronghorns are haunted by "the ghosts of predators past."[10] "Could it be," inquires Barbara Ehrenreich, "that human beings are similarly haunted by 'the ghosts of predators past'?"

> Darwin pondered the question when his two-year-old son developed a fear of large caged animals at the zoo. "Might we not suspect," the great amateur biologist wondered, "that the fears of children, which are quite independent of experience, are the inherited effects of real dangers . . . during savage times?"[11]

According to most developmental psychologists, Darwin's son simply *couldn't* have feared that the lions and tigers in London Zoo might kill him, because the idea of death only arises in a child's mind years later. But listening to children's answers to metaphysically loaded questions like "Where do dead people go?" is unlikely to get to the heart of the matter. Questions about death have confused great philosophers, so we should not expect it to be transparent to small children. But what if we investigate the matter differently and look at what these children make of realistic scenarios in which predators attack and kill prey? A different picture emerges. Even very young children are sensitive to the difference between death and sleep, understand that a dead animal cannot move, grow, or be conscious, and that the condition is irreversible. Research also shows that children regard dead animals as equivalent to meat, and children as young as one year old have an unconscious awareness of pursuit-evasion strategies.[12] As usual, Darwin was right on the money. There is an expanding body of research suggesting that the ancestral experience of predation had a decisive impact on the configuration of the human mind and that this ancient legacy continues to prime the pump of imagination and behavior. When we watch movies like *Snakes on a Plane*, we identify with the human characters and experience a small taste of the heart-pounding terror of a hunted animal from the safety of our cinema seats. What makes this happen? The most likely explanation is that the human mind comes equipped with a *predator detection module* that is switched on by images of dangerous animals. University of California anthropologist H. Clark Barrett points out that when we see a human being approaching us, we are open to many possibilities for interaction. "Perhaps," writes Barrett, "the person wants to ask directions, to give you a message from a friend, to propose an exchange of goods, to ask you for a date, or to kill you." In contrast, when we are approaching by a lion "all that matters is that he might attack."[13] When the predator-detection module is triggered, the result is simple and unambiguous: we either fight or flee.

In war, the soldier's predator-detection module can be switched on by other human beings. When this occurs, the enemy is no longer experienced as human. He is perceived as a dangerous beast that must be killed. The soldier, especially if isolated from his comrades, may feel that a relentless predator is stalking him. Here is how Guy Sajer describes the experience in his harrowing memoir of life on the Eastern Front during World War II:

> It is difficult even to try to remember moments during which nothing is considered, foreseen, or understood, where there is nothing under a steel helmet but an astonishingly empty head and a pair of eyes which translate nothing more than would the eyes of an animal facing mortal danger.[14]

So far, a few political cartoons and posters have had to bear the whole evidential weight of my argument. But these are examples of political propaganda, and propaganda does not give us any insight into the psychology of combat. This is a reasonable objection, but it is not a sound one. In the first place, we would still need to explain *why* images of predators are so effective in fanning the flames of war. The obvious answer is that the human mind naturally associates dangerous human beings with large, dangerous animals—that this sort of propaganda works because it takes advantage of a preexisting mental disposition to perceive enemies as ravenous man-beasts. Throughout history, and across the globe, evil is personified by human figures with nonhuman characteristics. We find it in Bosch's demons in *The Temptation of Saint Anthony,* in Japanese *oni* masks, and in the form of the ancient Mesopotamian demon Pazuzu, who combined a human form with a lion's head, an eagle's talons, and a scorpion's tail, and in the Minotaur, the Sphinx, and the gorgons of ancient Greece. In Homer's *Iliad,* acts of war are often likened to attacks by dangerous animals. When Patroclos is slain by Hector, he is "like a wild boar felled by a lion."[15] The warriors surrounding Diomedes are described as "furious as roaring lions or wild boars."[16] Patroclos "pounced on Cibriones like a lion that ravages the fold."[17] When the Achaeans flee from a surprise attack by the Trojans, it is "like a herd of cattle or a great flock of sheep chased by a couple of wild beasts in the murk of night."[18] The scene in the Danäans turn on their Trojan attackers could hardly be more explicit.

> The rest of the Danäans pounced on their enemies as ferocious wolves pick out lambs and kids from the flock and pounce on them . . . they tear the poor timid creatures to pieces at sight.[19]

The human imagination glides easily from human violence to predatory attacks by animals. Moving from the eighth century B.C. to the twentieth century A.D. we find strikingly similar imagery. Kaiser Wilhelm II, who ruled Germany during World War I, was described in the British press a "wild animal," "the beast of Berlin," and the "mad dog of Europe."[20] "Beast" and "mad dog" are popular epithets for enemies. Hitler, too, was the "beast of Berlin" and Saddam Hussein was the "beast of Baghdad." Ronald Reagan called Libya's Colonel Qaddafi "that mad dog of the Middle East."[21] Israeli premier Menachem Begin characterized his Palestinian enemies as "beasts walking on two legs."[22] Nazi propagandists depicted Jews as "beasts of prey" that the German people must destroy in self-defense. According to one SS pamphlet, the Jew "looks human . . . but his spirit is lower than that of an animal. A terrible chaos runs rampant in this creature, an awful urge for destruction, primitive desires, unparalleled evil, a monster, subhuman." More recently, Paul Wolfowitz, then the U.S. deputy defense secretary, remarked of the war on terror that "While we'll try to find every snake in the swamp, the essence of the strategy is draining the swamp." How often since then have we heard it said that mosques and madrasas are "breeding grounds" for terrorists, who must be "smoked out" of their hiding places (often, by the way, assumed to be caves) and "hunted down"?[23]

Second, fighting men on the ground often invoke hunting imagery, and it is implausible to put this down to the effects of political propaganda. Many soldiers described World War I as a grand hunting expedition. For example, British war hero Frances Grenfell's last words at Ypres were "Hounds are fairly running!" and Ernst Jünger described the Highlanders whom his men massacred as "hunted game."[24] According to Sergeant J. A. Caw, "for excitement, man hunting has all other kinds of hunting beat a mile."[25] Much the same was true in World War II, particularly in the Pacific Theater, where physical differences made the Japanese easier to demonize. American soldiers embarking on combat missions were often sent off with the words "Good luck and good hunting." "Sometimes the imagery chosen was of a general nature," writes historian John W. Dower, "of marines . . . shooting . . . predatory animals."

> Sometimes the hunt was violent. General Slim divided the decisive battle of Imphal-Kohima into four stages, the last of which was pursuit—"when the Japanese broke, and, snarling and snapping, were hunted from the field." At another point Slim recalled that "relentlessly we would hunt them down and when, desperate and rabid, they turned at bay, we killed them."[26]

Americans portrayed the Japanese as monkeys, gorillas, or mad dogs, while the Japanese referred to the Americans and the British as "wild beasts" and sometimes represented Roosevelt and Churchill as demonic beings with nonhuman hindquarters and small horns protruding from their temples.[27] World War II snipers described themselves as "big-game hunters" shooting at "soft-skinned dangerous game," and British fighter pilots used the same analogy.[28] "Even at sea," writes Joanna Bourke, "a ship might be described as 'the relentless hunter speeding after its prey' or being 'dead on the tail of our quarry.'"[29] During the war in Yugoslavia, Serbian killers were described as "beasts in human form," "bearded animals on two legs."[30] Veterans' accounts of the Vietnam War are full of hunting imagery. U.S. marine sniper James Hebron put it with startling frankness, remarking on "that sense of power, of looking down the barrel of a rifle at someone and saying, 'Wow, I can drill this guy.' . . . It's the throw of the hat, it's the thrill of the hunt."[31]

Images of predators and prey sometimes crop up in genocidal wars. For example, L. Frank Baum (author of *The Wizard of Oz*), referring to the remaining American Indians, recommended in 1891 that the United States government "wipe these untamed and untamable creatures from the face of the earth."[32] German colonists called the Herrero of Namibia fierce baboons, smoothing the way for the first genocide of the twentieth century. During the Rwandan genocide of 1993, Hutus described Tutsis as having had tails and pointed ears.[33] As we will see, though, wars of extermination typically draw their inspiration from a different set of psychological forces.

What is going on here? It seems to be something like this. Do you remember the peoplemaking module described in chapter 9, the brain system that causes us to perceive people as *people*? In combat, this module is sometimes switched off, and something more primitive takes its place. When this occurs, the enemy is no longer perceived as a human being. Instead, images of predators, of evil, devouring monsters, are awakened in the soldier's mind as ancient cognitive mechanism swing into action as quick and dirty fight-or-flight responses replace reflection, compassion, and humanity. When this unconscious, automatic process occurs in the heat of battle, it becomes possible for men to kill freely and even enjoy the act of killing.

SOUVENIRS OF SLAUGHTER

Hunters sometimes return home with trophies from the chase: a tiger-skin rug, a buck's head, or a lucky rabbit's foot. So it is not surprising that warriors also bring home trophies: body parts taken from the people whom they have killed. The custom of trophy taking goes back to prehistory. Heads were very popular items, and were taken by warriors in New Guinea, Oceana, North America, South America, Africa, and Stone Age Europe.[34]

> Less common trophies taken by tribes in various areas of the world included hands, genitals, teeth, and the long bones of the arms or legs. These long bones were made into flutes in South America and New Zealand. Several chiefdoms in Colombia kept the entire skins of dead enemies. Often the women who accompanied their men into the battlefield flayed the victims. One group even stuffed these trophy skins, modeled the features of the victims in wax on their skulls, place weapons in their hands, and set the reassembled trophy "in places of honor on special benches and tables within their household."[35]

In the Bible we are told that King David and his soldiers killed two hundred Philistines and cut off their foreskins, and there are also references in the Good Book to postmortem decapitations.[36] Herodotus recounts that the ancient Scythian warriors not only scalped their victims but also sewed the scalps together to make cloaks and skinned their arms and hands (complete with fingernails) to make quivers for their arrows.[37] The Aztecs had skull racks that could hold up to a thousand heads.[38] Wartime trophy collecting is not just a relic of a bygone age. During World War II, U.S. servicemen in the Pacific "harvested" gold teeth, severed hands from corpses, and took ears, bones, scalps, skulls, and even women's breasts.[39] War correspondent Edgar L. Jones wrote in the *Atlantic Monthly* that soldiers "boiled the flesh off enemy skulls to make table ornaments for sweethearts, or carved their bones into letter openers."[40] Although the soldiers who indulged in these grisly activities were a minority, the fact that customs officials in Hawaii routinely asked servicemen departing from the Pacific islands whether they were carrying any bones in their luggage suggests that trophy collecting was not all that rare.[41]

American and Australian soldiers in Vietnam also collected body parts, and some GIs strung ears or fingers around their necks, calling them "love beads" to parody the peace movement back home.[42] One marine recounted:

> We used to cut their ears off. . . . If a guy would have a necklace of ears, he was a good killer, a good trooper. It was encouraged to cut ears off, to cut the nose off, to cut the guy's penis off. A female, you cut her breasts off. It was encouraged to do these things. The officers expected you to do it or there was something wrong with you.[43]

These acts are so outrageously grotesque that it is tempting to think that their perpetrators must have been deranged. How can a normal man bring himself to sever body parts to keep as souvenirs? How can an ordinary, decent human being carve a human thigh bone into a pen holder, or slice through the cheek of a still-living soldier to pry out his gold teeth? The answer seems clear. Psychologically speaking, the victims are no longer human beings, and cutting off their fingers or ears is of no greater moral consequence than taking the head of a bear to mount over the fireplace.

PREY

If ancestral memories of fearsome predators were the only driving forces behind demonization of the enemy in war, my story would finish here. While it is partially true that "the emotions we bring to war are derived . . . from a primal battle that the entire human species might easily have lost," this is far from being the whole story. This is clear from a letter written over a hundred years ago by a young soldier serving in the Philippines, to his folks back home: "Our fighting blood was up, and we all wanted to kill 'niggers.' . . . This shooting human beings beats rabbit hunting all to pieces."[44] Here war is a hunt, but not a hunt for deadly predators. What are we to make of this?

The association between hunting for food and killing members of our own species probably goes back to our common ancestor the chimpanzee. Like humans, chimpanzees also go on hunting expeditions to kill and eat other animals (they especially like to hunt red colobus monkeys, but they also kill and eat other animals, including young antelopes and pigs). As with humans, cooperative hunting is an overwhelmingly masculine activity—females make less than 10 percent of all kills, and males sometimes give females meat in exchange for sex. Many primatologists believe that there is an evolutionary connection between chimpanzee intergroup violence and hunting behavior, although it is unclear whether raiding emerged from cooperative hunting, or vice versa. In any case, archeological remains show that our ancestors were hunting animals, both small and large, at least one and a half million years ago.[45] There is also evidence suggesting that prehistoric warriors may have hunted human beings for food. Archeologists have found heaps of human bones scarred with the distinctive cut and scrape marks produced when an animal is butchered, and discoveries of smashed skulls and thigh bones may be mute testimony to our ancestors' penchant for gorging themselves on brains and marrow. A spectacular find at Cowboy Wash, Colorado, of fossilized human feces containing traces of human muscle protein provides definitive evidence of early cannibalism, although it is not clear whether this occurred in the context of war (eating the dead may have been a funerary custom).[46]

A taste for hunting was bred into our species, and persists to this day. In the developed world, hunting is no longer a source of food, but the hunting impulse gets expressed in hunting for "sport." Even those who find killing animals distasteful satisfy these urges in other ways (for instance, by reading detective fiction, enjoying movies and computer games involving themes of pursuit and capture, and even bargain hunting). Children the world over are attracted to games like hide-and-seek and tag, which hone both hunting and predator-evasion skills, and to games involving target practice. All of this speaks to the existence of a deeply ingrained cognitive system specialized for the identification and capture of prey. Some cultures have explicitly recognized that warriors must become predators to do their job and help the process along by ritualistic means. Viking men were magically transformed into bears before becoming elite warriors, a process that was called going "berserk" ("dressed in bear skin"). The berserker would then attack and bite other people.[47] Tahitian warriors had to mimic wild dogs.[48] The notion that a being can have human form but a predator's essence is also expressed by the ubiquitous werewolf legend. The word "werewolf" comes from the Old English for "man-wolf." It is found all over the world, with local variations. In parts of Africa, for instance, there are said to be *bouda* or were-hyenas, and the Aztecs had their *nahuales* or were-jaguars. Written accounts of werewolves go back to the Greek legend of King Lycaon of Arcadia, who was transformed into a wolf as punishment for eating human flesh, but the idea is almost certainly much older—perhaps going back to Stone Age shamanistic beliefs. As hunters, *we* are the alpha predators, and nature has given us the cognitive software to make this happen.

Perceiving the enemy as a game animal is another way to sidestep the taboo against killing human beings. When this happens, the enemy is dehumanized without being demonized: they are not monsters, but neither are they people. They are innocuous game that can be gunned down for sport. It is easy to discover this attitude toward the enemy in the literature of war. During World War I, for example, Australian servicemen compared shooting enemy combatants with "potting kangaroos in the bush," while Americans and Brits spoke of shooting "human rabbits" and birds. In both world wars, infantrymen were often compared to poachers, and British soldiers in World War II compared the pursuit of Japanese soldiers with fox hunting.[49] Attacking huge, lumbering tanks was "boar hunting," as imagination transformed metal and rivets into animal flesh and blood.[50]

The American media often represented war in the Pacific as a pleasant hunting excursion.

> Very frequently . . . the hunt was pastoral, almost lazy, and the quarry small and easy. A cover story in *Life*, showing GIs walking through the jungle with rifles ready, looking for Japanese snipers, explained that "like many of their comrades they were hunting for Japs, just as they used to go after small game in the woods back home." A 1943 book giving a firsthand account of the combat explained that "every time you hit a Jap [with rifle fire] he jumps like a rabbit." The Battle of the Philippine Sea . . . became immortalized as "the great turkey shoot." "Duck hunting" was another popular figure of speech. . . . Killing Japanese reminded others of shooting quail. "Tanks were used to flush the Japanese out of the grass," a journalist reported from Guadalcanal, "and when they are flushed, they are shot down like running quail."[51]

This way of conceiving of war is implicitly cannibalistic. Ducks, turkey, rabbits, quail, deer, boar, and even kangaroo are hunted for food as well as for sport. This is sometimes made quite explicit, as when Admiral William "Bull" Halsey spoke of a naval operation against Japan as going "to get some more monkey meat."[52] Historian John Dower tells us:

> *Leatherneck* [a USMC publication] ran an unusually wild comic strip in which an unkempt white soldier—infuriated when a canteen of liquor is shot out of his hands by the "slant-eyed jerks" and "jaundiced baboons"—plunges into the jungle and emerges with four dead creatures with monkey's bodies and Japanese faces, tied by their tails and hanging from his shoulders. "They're a bit undersized," he exclaims, "but I got four of 'em!!"[53]

Japanese also conceived of their enemies as prey. One veteran, recalling how he and his buddies raped and then shot to death Chinese women in Nanking, remarked, "Perhaps when we were raping her we looked at her as a woman, but when we killed her, we just thought of her as something like a pig."[54] In Vietnam, servicemen wore jacket patches emblazoned with the words "Viet Cong Hunting Club" (these are still made and sold to fans of Vietnam memorabilia), and killing Viet Cong was said to be "just like hunting for deer." Even the well-worn clichés of war are often redolent of the chase. During the Vietnam War, Lyndon Johnson entreated U.S. troops to "nail that coonskin to the wall." "Where else," wrote R. B. Anderson in a chilling essay on his time in Vietnam, "could you divide your time between hunting the ultimate big game and partying at 'the ville'?"[55]

More recently, U.S. airmen described the bombing and strafing of Iraqi troops on the road to Basra in the First Gulf War as a "turkey shoot" (the Iraqis were said to be "sitting ducks"). In much the same spirit, in the opening stages of the war, Pentagon officials described the 2003 invasion of Iraq as "like shooting fish in a barrel" and (once again) as a "turkey shoot."[56]

Examples of hunting imagery in war could be multiplied almost ad infinitum, but I think that these are sufficient to convey the point. By conceiving of the enemy as predator or as prey, soldiers are able to release the killer instinct that enables them to do their job. But this is still not the full story. There is still a third form of dehumanization to consider before our survey is complete.

POLLUTION

Predators were not the only natural enemy of prehistoric human beings. Our ancestors had to confront another foe that was far more insidious and difficult to fight. These were parasites. Some are macroscopic—ticks, lice, worms, and so on—but most of them are invisible to the naked eye. These diminutive monsters are even more deadly than the larger predators. You can spot a lion coming, and take evasive action. You might even be able to drive it away or kill it. But how do you deal with a microscopic assassin? I mention this because the metaphor of the enemy as a disease, or as a carrier of disease, often occurs in war, especially in wars of extermination. Here is an example from the mouth of Adolf Hitler:

> The discovery of the Jewish virus is one of the greatest revolutions that has taken place in the world. The battle in which we are engaged today is the same sort as the battle waged, during the last century, by Pasteur and Koch. How many diseases have their origin in the Jewish virus! We shall regain our health only by eliminating the Jew.[57]

Hitler was not just using a rhetorical gimmick. He was in the grip of powerful feelings with profound biological underpinnings. To understand this properly, we must first have a brief look at the impact that parasites have had on the evolution of the human mind.

Parasitic organisms have always been our most deadly enemies. *Yersinia pestis* is a tiny bacterium that causes bubonic plague, a disease that killed around 25 *million* Europeans in the fourteenth century—more than half the population Europe. It is so virulent that even a single bacterium entering the bloodstream can cause infection, and the disease kills almost everyone that it infects. Although we often think of the bubonic plague as a phenomenon of the Middle Ages, the last major outbreak occurred in the second half of the nineteenth century and took the lives of over 12 million people, mainly in China and India. *Vibrio cholerae*, the bacterium that causes cholera, is transmitted by contaminated water. It causes diarrhea and vomiting and kills its victims by causing them to loose massive quantities of bodily fluids. *Variola major* causes smallpox. Once a scourge of mankind, it survives today only in American and Russian military laboratories, where it is preserved because of its potential utility as a biological weapon. The last outbreak, in 1967, killed two million people. The virus causing the Spanish flu is the deadliest on record: it snuffed out around 50 million human lives in only a year and a half.[58] Malaria, which is caused by blood-borne parasites transmitted by mosquito bites, kills up to three million people every year. Then there is AIDS, typhoid fever, sleeping sickness, syphilis, Ebola, leprosy, yellow fever, tuberculosis, measles, Lhasa fever . . . Not to mention macroparasites like eighty-foot-long tapeworms, brain worms, guinea worms, trichina worms, and many other nightmarish creatures.

Although most of these parasites are microscopically small, they are not entirely invisible because many of the signs and symptoms indicating their presence—signs like oozing sores, swellings, discoloration of the skin, and fetid odors—are well within the sensory range of most animals. Furthermore, they get transmitted by substances like feces, vomit, blood, pus, spittle, stagnant water, and decaying flesh, as well as by vermin like rats, flies, ticks, and mosquitoes, all of which are easily detectable. Like predators, parasites have been a recurrent selection pressure for a very long time, and we should expect the human brain to have evolved specialized systems that respond to their presence.[59] Of course, these cognitive systems could not detect the microorganisms themselves. They would have to be sensitive to the outward signs of infection, and to the substances and animals that are most likely to transmit it. In this chapter I will make the case that evolved mechanisms for parasite avoidance play a crucial role in the psychology of war.

REVULSION

In *The Anatomy of Disgust*, University of Michigan law professor William Ian Miller tells an instructive story about an episode in the life of the fourteenth-century saint Catherine of Siena. It goes like this:

> A sick nun had a cancer on her breast that put forth such an awful stench that no one would attend her in her sickness. Catherine volunteered to care for her but one day "when she was about to open the sore to dress it, there

> came out such an horrible stench, that she could hardly bear it, but that she must needs vomit." Catherine was upset at her own squeamishness and resolved to overcome it. "She bowed down and held her mouth and nose over the sore so long until at the length it seemed that she had comforted her stomach and quite overcome the squeamishness that she felt before."

Unfortunately for Catherine, the effect of this cure did not last. Satan intervened to make her nauseous again, and she vomited. But this time she resorted to stronger medicine.

> Catherine, seeing that it was the work of the serpent, nonetheless took "an earnest displeasure against her own flesh," which she addressed as follows: "I shall make thee not only endure the savor of it, but also to receive it within thee. With that she took all the washing of the sore, together with the corrupt matter and filth; and going aside put it all into a cup, and drank it up lustily. And in so doing, she overcame at one time, both the squeamishness of her own stomach and malice of the Devil." That night Christ came to her in a dream, and in reward for drinking pus . . . he drew her mouth to the wound on his side and let her drink her fill.[60]

Pretty revolting, isn't it? Reading the vignette, we identify with Catherine and imagine ourselves drinking the bowl of malodorous pus. To understand why this has such a stomach-churning effect, we must consider the evolutionary psychology of the feeling of disgust. We all must eat to live, and because eating is essential to life, nature has made it pleasurable. We do not love the taste of a sweet peach or a succulent steak because these things are *intrinsically* delicious. In fact, there is no such thing as intrinsic deliciousness (if you were a dung beetle or a maggot, you would have a very different set of culinary preferences). We find some items tasty and others revolting because we are the descendents of women and men who benefited from the former and avoided the latter. Imagine an ancestral population consisting of some that enjoyed the taste of fresh meat and others who savored the taste of rotting, maggot-infested meat. Which group do you think would have the best survival prospects? The general formula is delightfully simple. Forget about the puritanical notion that whatever is enjoyable can't be too good for you: the apostles of Darwin preach a more hedonistic gospel. Generally speaking, present-day human beings enjoy the taste of foods similar to those that benefited their remote ancestors, and are repelled by the taste of foods that were harmful to their ancestors. This is not to deny the existence of significant cultural and individual variations in food preferences, but it is the broad, near-universal tastes that concern us here.

Disgust is an adaptive response to eating something harmful. It causes retching: the body's attempt to expel the harmful object. A disgusting object is *offensive*; although it is not necessarily bad-tasting (do you really know what a bowl of pus actually *tastes* like?). Things are disgusting because of what they *are*, not because of how the smell or taste (we are disgusted by flavors only by association: tastes are disgusting only because they remind us of disgusting things).[61] The disgusting character of an object is contagious: it contaminates edible things with which it comes in contact and renders them inedible. A person who does not like the taste of cilantro is unlikely to object if one cilantro leaf falls into a cauldron of stew, but how different the reaction would be if a small cockroach fell into the pot! Disgust reactions are not confined to foodstuff. They extend to bodily fluids, excretory products, certain sexual practices, gore, decaying objects, and corpses. Intimate contact with strangers also elicits disgust—for example, wearing their undergarments or using their toothbrushes.

People the whole world over instinctively approach this domain with a special set of assumptions. Even those with no understanding of the role of microscopic organisms in contagion realize that the danger posed by "unclean" things does not have to be visible—it is typically thought that the contaminated object transmits its filthy *essence* to anything that it comes into contact with. It follows that even brief contact causes a person or thing to become *wholly* contaminated and that this effect may be irreversible. (According to one Zoroastrian text, anyone who has had contact with a corpse is polluted "to the ends of his nails and unclean for ever and ever.") Finally, it is thought that any form of contact with the unclean object can cause contamination, although consuming it is an especially potent mode of contagion.[62]

Some investigators have explained these phenomena as manifestations of our fear of mortality or as a forlorn attempt to reject our "animal nature," but I do not think that anybody with a biological slant on human behavior can be satisfied with these attempts. There is a much simpler and more obvious evolutionary story to be told about the origins of disgust, namely that *feelings of disgust are evoked by potential sources of infection*. Bodily fluids, feces, blood and gore, and, of course, corpses are all major vectors of transmission. Intimate contact with strangers, including sexual contact, is also a potent source of infection. "A person's tongue in your mouth," writes William Ian Miller, "could be experienced as a pleasure or as a most repulsive and nauseating intrusion depending on the state of relations that exist . . . between you and the person."[63] When human beings lived in small isolated groups, encounters with strangers were potentially threatening because you might not have acquired a resistance to the germs carried by the outsider. It was only a few hundred years ago that this fate befell native Americans, who were ravaged by the germs brought by European colonists and conquistadores. Taking this a little farther, it is not too far-fetched to think that vulnerability to disease may underpin our hostility toward strangers. The ancient Greeks believed foreigners possessed an "evil eye" that could cause illness, and psychological studies have shown that people who feel most vulnerable to disease also tend to be xenophobic.[64] The fear of disease may even contribute to racial prejudice. Our ancestors must have been exposed to very little human variation. An encounter with someone with a different skin color or other unusual physical features would have been very startling, and would have more often than not indicated that the person was diseased. If this is hard to grasp, it is because we are so accustomed to human diversity. But imagine getting up in the morning, looking in the bathroom mirror, and noticing that your skin was not its normal color. I bet that your *first* reaction would be to think that you are suffering from a disease. Even though today we know better, the ancient tendency may continue to lurk in the archaic depths of the modern mind.

The robustness of disgust reactions and their universality suggest they are deeply rooted in human nature, and that our ancestors evolved a mental module specifically for dealing with the risks of parasitic infection. Like the modules for detecting and dealing with predators and prey, the antiparasite module plays an extremely important role in the dehumanizing of enemies in war.

EXTERMINATION

When the antiparasite module is activated in war, we perceive our enemies as unclean. We equate them with disease, or with vectors of disease such as lice, flies, and rats. We do not hunt rats for sport, and do not mount their heads on the wall as trophies, nor do we eat creature that we consider to be filthy. We try to wipe them out, to exterminate them completely. It follows that when the antiparasite module is activated and turned against fellow human beings, the stage is set for genocide.

In the eleventh century, Pope Urban II whipped up support for the First Crusade in a speech that graphically portrayed the Muslim stewards of Jerusalem as unclean. He proclaimed that the Christian holy places were held by "unclean nations" who "destroy the altars, after having defiled them with their uncleanness. They circumcise the Christians, and the blood of the circumcision they either spread upon the altars or pour into the vases of the baptismal font."[65] Centuries later, during the colonization of America, Indians were portrayed as carriers of disease. This is chillingly illustrated by Captain Wait Winthrop's 1675 poem proposing a policy of extermination:

But humbled be, and thou shalt see these Indians soon will dy.
A Swarm of Flies, they may arise, a Nation to Annoy,
Yea Rats and Mice, or Swarms of Lice a Nation may destroy.[66]

Almost two centuries later, whites were still using the same kind of imagery. One popular expression from the mid–nineteenth century is "A nit would make a louse" (or "Nits make lice") A nit is a louse egg, and "nits make lice" was an oblique way of saying that Native American children should be slaughtered along with their parents.

The Jewish Holocaust provides a particularly gruesome example of extreme contagion panic and its deadly effects. Christians had long portrayed European Jews as vectors of infection. This dates back at least to the Middle Ages, when the Jews were accused of spreading the bubonic plague.[67] However, Hitler and the National Socialist Party whipped up this fear to an unparalleled pitch. Holocaust historian Lucy Dawidowitz informs us:

> The vileness of the Jew . . . Hitler declared, had permeated nearly every aspect of modern society. Over and over again he kept describing the Jews in terms of filth and disease. "If the Jews were alone in this world, they would stifle in filth and offal." Jews, he asserted, were at the center of every abscess, were "germ carriers," poisoning the blood of others, but preserving their own. The Jews were, Hitler said, "a ferment of decomposition."[68]

The Nazis repeated this rhetoric obsessively, and regarded it with deadly seriousness. As Hitler put it in *Mein Kampf*: "Was there any form of filth or profligacy . . . without at least one Jew involved in it? If you cut even cautiously into such an abscess, you found, like a maggot in a rotting body . . . a little Jew."[69] Other Nazis followed suit. Übelhör, governor of the Kalisz-Lodz District of Poland, compared the extermination of the Jews to burning out the bubonic plague. Reichsführer-SS Heinrich Himmler, a leading architect of the Final Solution, declared of the Jews, "We have exterminated a germ."[70] The notorious 1940 propaganda film *The Eternal Jew* compared this "parasitic race" to teeming swarms of rats. The narrator states:

> In this way, they [the rats] spread disease, plaguc, leprosy, typhoid fever, cholera, dysentery, and so on. They are cunning, cowardly, and cruel, and are found mostly in large packs. Among the animals, they represent the rudiment of an insidious and underground destruction—just like the Jews among human beings.[71]

The Nazis were not the only ones to associate their enemies with disease-carrying rats. John Dower explains:

> Allied forces picking apart routed Japanese troops were described like being "terriers onto rats." Reporters spoke of the "beady little eyes" of critically wounded Japanese, and Japanese soldiers . . . were likened to rats caught in their hole. . . . In February, 1945, when the Battle of Iwo Jima was taking place, the *New York Times* ran an illustrated advertisement by a US chemical company showing a GI blasting a path "through stubborn Jap defenses" with a flamethrower. The ad bore the heading "Clearing out a rat's nest." . . . Many marines actually went into battle with the legend "Rodent Exterminator" stenciled on their helmets.[72]

The Japanese were also thought of as lice. *Leatherneck* portrayed the Japanese as a grotesque parasite, called "Louseous Japanicus," and explained that the marines were responsible for "the giant task of extermination" and that "the breeding grounds around the Tokyo area . . . must be completely annihilated."[73]

The genocidal Khmer Rouge regime in Cambodia referred to its enemies as "microbes," "pests buried within," and traitors "boring in," and Pol Pot exhorted his disciples to kill "ugly microbes." Slogans like "What is infected must be cut out" and "What is rotten must be removed" urged party members to eliminate corrupt individuals from "the brotherhood of the pure."[74] Upper-class Nigerians thought of their Ibo neighbors as vermin, and the French in Algeria called Muslims "rats." The dictatorial Balinese New Order government implemented a "clean environment" policy. The relatives of people alleged to be communists were "infected" by "political uncleanliness"[75] Rwandan Hutus depicted Tutsis as "cockroaches" and their genocide as a "big cleanup."[76] As one informant mentioned to his French interviewer, "We called them cockroaches, an insect that chews up clothing and nests in it, so you have to squash them hard to get rid of them."[77] During the Yugoslavian wars Bosnian Serbs were described as a "malignant disease" threatening to infect Europe. Serbian killers were described in the Croatian press as "bloodsuckers."[78] Ronald Reagan called Marxism a "virus" and vowed to battle against the "communist cancer."[79] United States military officers referred to the town of Fallujah as a "rat's nest" and a "cancer" just before the devastating assault in November 2004.[80]

When enemy is thought of as filth, war is conceived as a grand hygiene operation. The expression "ethnic cleansing" came into popular usage during this conflict, as when ethnic Albanians demanded an "ethnically clean" Kosovo.[81] The Nazis, who used the term "Cleansed of Jews" (*Judenrein*) for territories from which Jews had been eliminated. "Cleansing of borders" was similarly used under Stalin. Expressions like "purging," "mopping up," "cleaning out," and "wiping out" are so common in the rhetoric of war that we seldom give them much thought (for example, after the 2003 invasion of Iraq, U.S. troops were said to be "mopping up" the insurgency).[82]

In this chapter I have tried to give an account of how dehumanizing the enemy in warfare draws on ancient biological dispositions to overcome the problem posed by the taboo on killing members of our own species. To do this, particular mental modules are activated which cause the soldier to perceive his enemies as human in form but lacking a truly human essence. Linking back to the material presented in chapter 6, we can regard this as a form of self-deception. In a sense, the soldier must lie to himself about what he is doing. He is not spilling the blood of others, he is killing an evil beast, or shooting turkeys, or ridding the world of a terrible disease. To the extent that he can sustain this illusion, he can avoid psychological deterioration; to the extent that he cannot, he falls prey to the symptoms described in chapter 8. The evolutionary psychology of dehumanization is intuitively grasped by propagandists, who use it to inspire young men to do their bidding, and it also arises spontaneously in situations of mortal conflict. The scope of these illusions about "the Other" extends far beyond the battlefield. They may be even more prevalent behind the lines, where slogan-spouting politicians and flag-waving crowds get their pleasure on the cheap. They can bay for the enemy's blood without ever putting their lives on the line or paying the terrible spiritual price of killing.

11

HUMANITY LOST AND FOUND

> And so, to the end of history, murder shall breed murder, always in the name of right and honor and peace, until the gods are tired of blood and create a race that can understand.
>
> —GEORGE BERNARD SHAW, *CAESAR AND CLEOPATRA*

IN BOOKS LIKE THIS it is conventional to close with reflections on the future of war. In doing this, it is easy, or perhaps even conventional, to substitute platitudes for thinking. The future is hard to think about, just because it is the future, and war is hard to think about, for reasons that I hope I have made plain in this book. Together, they make an unappealing cocktail. I cannot make any confident pronouncements about the future of war. I certainly disagree with those who argue that because the sheer destructiveness of war has become so great, before long it will no longer be an option for humanity.[1] Such approaches assume that war is generally a rational endeavour, an approach that I think is misconceived. Taking my cues from the past, I am far from optimistic about the future.

Still, there may be cause for very cautious optimism. Although the dice appear to be loaded in war's favor, the outcome is not inevitable. The psychological alchemy that transmutes human beings into nonhuman beings plays an important role in enabling soldiers to overcome their inhibitions against killing and affords them protection from the horrendous psychological trauma of war. Cast your mind back to William Manchester's description of gunning down a trapped Japanese sniper, an act that caused him psychological pain that probably lasted a lifetime. If he had perceived the sniper as a rat in human form, he would have been spared this agony. Father Agostino Gemelli, a chaplain on the Italian front during World War I, was keenly aware of the psychological changes that soldiers *must* undergo to avoid madness. At first, he could not comprehend what it was that enabled Italian soldiers, many of whom were uncultured and ignorant of the cause that they were allegedly fighting for, to take great risks and perform heroic feats. But he soon realized what had happened to them: they "ceased to be men" for the duration of the action. When it was all over and they returned to their line, they broke into tears or collapsed, signaling that another change had taken place: "Human nature took over again." The men regained the human self that they had lost when they went into battle.

If this analysis is anywhere near correct, then our best hope of stopping war is stopping this kind of self-deception, or at least becoming intolerant of it. If we do not take refuge in illusion, we will find it much more difficult to go to war. Abandoning humanity, even for a short time, is a risk, for there is never any guarantee that one will regain it. The longer a man remains in battle, the more difficult it becomes for him to rediscover his humanity, and the greater the danger of breakdown. As with every drug, the effects of self-deception come at a price. The transformations that make war possible can unleash terrible forces within the men that undergo them. War offers unique and forbidden pleasures and satisfactions, which we can glimpse in the writings of men who are honest and reflective enough to give a true account of them. These letters and memoirs spell out thoughts and emotions that sane, civilized human beings are not supposed to have. Philosopher-soldier J. Glenn Grey describes some of these thoughts in his intensely moving memoir of the World War II, *The Warriors*. "Happiness is doubtless the wrong word for the satisfaction that men experience when they are possessed by the lust to destroy and kill their kind," he writes.

> Most men would never admit that they enjoy killing, and there are a great many who do not. On the other hand thousands of youths who never suspected the presence of such an impulse in themselves have learned in military life the mad excitement of destroying . . . the delight in destruction slumbering in most of us. . . . When soldiers step over the line that separates self-defense from fighting for its own sake, as it is so easy for them to do, they experience something that stirs deep chords in their being.[2]

Henri de Man expressed similar feelings. De Man fought in World War I and later became leader of the Belgian Socialist Party. A cultivated person, he thought himself immune to what he called the "intoxication" of slaughter—until the moment he secured his first direct artillery hit on an enemy position. As he watched pieces of men's bodies fly up into the air and listened to the screams of the wounded, de Man experienced such extreme pleasure that he *wept with joy.* Others describe their experience of war in unmistakably erotic terms. For Vietnam veteran Philip Caputo it was "like getting screwed the first time," an "ache as profound as the ache of orgasm." Former *New York Times* foreign correspondent Chris Hedges also points to the bizarre sexuality of the war zone:

> The seductiveness of violence . . . the god-like empowerment over other human lives and the drug of war combine, like the ecstasy of erotic love, to let our senses command our bodies. Killing unleashes within us dark undercurrents that see us desecrate and whip ourselves into greater orgies of destruction.[3]

The act of killing is supremely rich in sensory input. Think of a lion bringing down a zebra. There is the taste of blood, and the smell of blood, the prey's helplessness, the struggle, and its shrill cry of distress. These sensations must have excited prehistoric humans, too, and with sexual rewards in the offing, we can add mounting erotic tension to the mix. War is simultaneously hideous and exciting. Once in the grip of the illusions of war, many men take pleasure in killing. This is why the sight of gore and the screams of pain made Henri de Man so euphoric, and it is why Roderick Chisholm, a World War II fighter pilot, described shooting down enemy planes as "sweet and very intoxicating." It is why Winston Churchill could write to Violet Asquith, with amazing frankness, "I *love* this war. I know it's smashing & shattering lives of thousands every moment—& yet—I enjoy every second of it."[4] And this is true even though war often destroys the bodies and souls of men who love it.

War is both intensely horrible and exquisitely pleasurable. It is horrible because of the danger and suffering that soldiers and civilians endure, and the unavoidable guilt that comes with killing. It is pleasurable because—like all pleasures—it is something that benefited our ancient ancestors who were victors in the bloody struggle for resources. The joy of war is the joy of the hunt, of bringing down game, of ridding the world of a man-eating monster or obliterating a plague.

Robert E. Lee was right when he remarked, "It is well that war is so terrible, or we should grow too fond of it." War is terrible for warriors, and for civilians. But it is not necessarily terrible for those who send men to war, who stand to gain wealth, power, or a place in heaven while paying little or no price. From their perspective, it is best that we remain asleep and, intoxicated by fantasies, step into the abyss. It is the double-sided character of our attitude toward war that makes this possible. Our relationship with killing is ambivalent, a compound of pleasure and aversion. Both are deeply rooted in human nature, and neither can be extirpated. If I am right, we will never stop men from enjoying war, and trying to do so is a fool's errand. The most that we can hope for, in the end, is for men to detest it more than they enjoy it, and the only way to shift that balance is to expose the self-deception that makes killing bearable. If we can do this, however incompletely, we will have accomplished something heroic indeed.

APPENDIX

A PARTIAL LIST OF DEMOCIDES COMMITTED DURING THE PAST 100 YEARS

> If there were a Last Judgment as Christians believe, how do you think our excuses would sound before that final tribunal?
>
> —BERTRAND RUSSELL, *UNPOPULAR ESSAYS*

- Eight million residents of the Congo Free State killed by the Belgians between 1877 and 1908.
- Sixty-five thousand Namibian Herero killed by the Germans between 1904 and 1907.
- One and a half million Armenian Christians killed by Muslim Turks in 1915–16.
- Five million Ukranians killed in 1931–32 by the Soviet Union's perpetration of famine.
- Over four million Soviet citizens killed by their own government in the Great Terror of 1937–1938.
- Over three hundred thousand Chinese residents of the city of Nanking killed by the Japanese in 1937.
- Eleven million Jews, Roma, Poles, homosexuals, and others killed by the Germans during the 1940s.
- Over two hundred fifty thousand Muslims, Serbian Orthodox Christians, Roma, and others killed in death camps

run by the Roman Catholic Ustashi regime in Croatia between 1941 and 1945.

- More than two hundred thousand Muslims killed by the French in the 1954–62 war for Algerian independence.
- Around one million Indonesians killed by their own government in 1965–66.
- One million seven hundred thousand Cambodians killed by the Khmer Rouge during the 1970s.
- Roughly two and a half million people, mainly Hindus, killed by the Muslim Pakistani army in East Bengal in 1971.
- Around one hundred fifty thousand Hutus killed by Tutsis in Burundi in 1972.
- Around two hundred thousand Maya killed by the government of Guatemala between 1970 and 1996.
- Two hundred thousand Muslims killed by Serbian Orthodox Christians in Bosnia-Herzegovina during the 1990s.
- Close to one million Tutsi killed by the Hutu majority in Rwanda in 1994.
- Two hundred thousand Roman Catholics in East Timor killed by the Muslim Indonesian occupation force between 1975 and 1999.
- An as yet undetermined number of Muslims killed by Serbian Orthodox Christians during the 1990s.
- Around two million black Sudanese killed in Darfur by the government of Sudan, which is ongoing at the time of writing.
- An undetermined number of Anuak killed by the government of Ethiopia, ongoing at the time of writing.

Other victims of twentieth-century genocides include the Bubi of Equatorial Guinea, the Dinka, Nuba, and Nuer of Sudan, the Isaak of Somalia, the Karimojong of Uganda, the San of Angola and Namibia, the Tuareg of Mali and Niger, the Tyua of Zimbabwe, the Atta of the Philippines, the Auyu of West Papua and Indonesia, the Dani of

Papua New Guinea, the Hmong of Laos, the Kurds of Iraq, the Nasioi of Papua New Guinea, the Tamil of Sri Lanka, the tribal peoples of Bangladesh, the Ache of Paraguay, the Arara, Ticuna, Nambiquara, and Yanomami of Brazil, the Cuiva, Nunak, and Paez of Colombia, the Mapuche of Chile, the Maya of Guatemala, and the Miskito of Nicaragua. Today's genocides and ethnocides often take place at the behest of multinational corporations eager to acquire resources, typically by dispossession and environmental degradation. These include oil interests in Ecuador, Burma, Nigeria, copper and cold mining in West Papua, farming in Tanzania, logging in Malaysia, and uranium mining in Australia.[1]

NOTES

PREFACE

1. Unless the context demands otherwise, I will use the word "soldier" as a synonym for "military combatant."
2. J. Ellis, *The Sharp End: The Fighting Man in World War II.* (New York: Charles Scribner's and Sons, 1980, p. 110) Ellis gives many similar examples.
3. D. Davidson, "Mental events," in W. G. Lycan, ed., *Mind and Cognition: An Anthology*, 2nd ed. (Malden, MA: Blackwell, 1999), 39.

1. A BAD-TASTE BUSINESS

1. J. Glover, *Humanity: A Moral History of the Twentieth Century* (New Haven, CT: Yale University Press, 1999). A. Thomson, *Smokescreen: The Media, the Censors, the Gulf War* (Tunbridge Wells: Laburnham, 1992).
2. M. Bell, *In Harm's Way: Reflections of a War-Zone Thug* (London: Hamish Hamilton, 1995), 214–215.
3. R. Holmes, *Acts of War: The Behavior of Men in Battle* (London: Free Press, 1985). C. Carrington, *Rudyard Kipling: His Life and Work* (London: Macmillan, 1955), cited in Holmes, *Acts of War,* 60. S. Moeller, *Shooting War* (New York: Basic Books, 1989). W. Frassanito, *Antietam: The Photographic Legacy of America's Bloodiest Day* (New York: Scribner's, 1978). F. Ray, "The case of the rearranged corpse," *Civil War Times,* October 1961, 19.
4. "General Sir John Hackett on World War II films," *Sunday Times Supplement,* March 20, 1983, cited in R. Holmes, *Acts of War,* 67.
5. M. Kukler, *Operation Barooom* (Gastonia, NC: Self-published, 1980). Joanna

Bourke discusses this subject in detail in *An Intimate History of Killing* (New York: Basic Books, 1999).

6. A. Russell, "I was protecting you from a madman, Bush tells America," *Telegraph*, October 10, 2003. R. Morse, "Bin Laden's relatives made a first-class escape," *San Francisco Chronicle*, November 9, 2001.
7. E. A. Zillmer, M. Harrower, B. A. Ritzler, and R. P. Archer, *The Quest for the Nazi Personality: A Psychological Investigation of Nazi War Criminals* (Hillsdale, NJ: Lawrence Erlbaum, 1995), 194.
8. M. Harrower, "Were Hitler's henchmen mad?" *Psychology Today*, July 1976, 76–80.
9. H. Hohne, *The Order of the Death's Head: The Story of Hitler's SS* (New York: Ballantine, 1971), 405. C. R. Browning, *Ordinary Men: Reserve Police Battalion 101 and the Final Solution in Poland* (New York: Ballantine, 1992). J. Waller, *Becoming Evil: How Ordinary People Commit Genocide and Mass Killing* (Oxford: Oxford University Press, 2002), 87. I do not mean to imply that all of the Nazi killers were ordinary citizens. The Dirlewanger brigade was a killing unit that was commanded by and consisted largely of criminals. See V. N. Dadrian, "The comparative aspects of the Armenian and Jewish cases of genocide: a sociohistoric perspective," in A. S. Rosenbaum, ed., *Is the Holocaust Unique? Perspectives on Comparative Genocide* (Boulder, CO: Westview Press, 2001).
10. P. Levi, *The Drowned and the Saved* (New York: Vintage, 1989). H. Arendt, *Eichmann in Jerusalem: A Report on the Banality of Evil* (New York: Penguin Books), 276. See also M. Shermer, *The Science of Good and Evil: Why People Chent, Gossip, Care, Share and Follow The Golden Rule.* (New York: Henry Holt, 2004).
11. M. Twain, "Man's place in the animal world," *Collected Tales, Sketches, Speeches and Essays, 1891–1910* (New York: Library of America), 210.
12. T. A. Walker, A *History of the Law of Nations*, vol. 1: *From the Earliest Times to the Peace of Westphalia, 1648* (Cambridge: Cambridge University Press, 1899), 124.
13. M. Gellhorn, *The Face of War* (New York: Atlantic Monthly Press, 1988), 2.
14. Cited in D. A. Grossman, *On Killing: The Psychological Cost of Learning to Kill in War and Society* (New York: Little Brown, 1995), 74–75.
15. G. Sajer, *The Forgotten Soldier* (Washington, DC: Brassey's, 2000), 184–185.
16. For a broad survey of evolutionary psychology, see D. Buss, *Evolutionary Psychology: The New Science of the Mind*, 2nd ed. (New York: Allyn & Bacon, 2003). For an intriguing hypothesis about the evolutionary roots of artistic creativity, see G. Miller, *The Mating Mind: How Sexual Choice Shaped the Evolution of Human Nature* (New York: Anchor, 2001).

17. T. Hobbes, *Leviathan* (London: Penguin, 1981), 184.
18. Ibid., 185.
19. D. Buss, *The Murderer Next Door: Why the Mind Is Designed to Kill* (New York: Penguin, 2005).
20. See the works of Norman Cohen, especially *The Pursuit of the Millennium: Revolutionary Millenarians and Mystical Anarchists of the Middle Ages* (Oxford: Oxford University Press, 1970); *Europe's Inner Demons: The Demonization of Christians in Medieval Christendom* (Chicago: University of Chicago Press, 2001); and *Warrant for Genocide: The Myth of the Jewish World Conspiracy and the Protocols of the Elders of Zion* (El Segundo, CA: Serif Publishing, 1996).
21. Cited in S. Keen, *Faces of the Enemy: Reflections of the Hostile Imagination*, 2nd ed. (San Francisco: HarperCollins, 1991), 30.
22. I. Brown, *Khomeini's Forgotten Sons: The Story of Iran's Boy Soldiers* (London: Grey Seal, 1990).
23. L. H. Keeley, *War Before Civilization* (Oxford: Oxford University Press, 1996).
24. C. Carpenter, "Aggressive behavioral systems," in R.L. Holloway, ed., *Primate Aggression, Territoriality and Xenophobia* (New York: Academic Press, 1974), 491–492.
25. For raiding, see N. Chagnon, (1968) "Yanomammo social organization and welfare," in M., Fried, M. Harris, and R. Murphey, eds., *War* (Garden City, NY: Natural History Press); N. Chagnon, *Yanomammo: The Fierce People* (New York: Holt, Rinehart & Winston, 1983); I. Hogbin, A *Guadalcanal Study: The Koaka Speakers* (New York: Holt, Rinehart & Winston, 1964). Keeley, *War Before Civilization*. It may be that ritualized battle evolved as a consequence of frequent lethal warfare. According to Hölldobler, something similar is found in the behavior of the honey-pot ant that has evolved forms ritualistic display to replace costly mandible-to-mandible combat. The ant armies perform fighting dances. The colony with the largest number of dancers wins the tournament, the queen of the smaller colony is then killed and the workers are incorporated as slaves. See B. Hölldobler, "Tournaments and slavery in a desert ant," *Science* 210 (1976): 912–914; B. Hölldobler, "Foraging and spatiotemporal territories in the honey ant *Myrmecocystus mimicus*," *Behavioral Ecology and Sociology* 9 no. 4 (1981): 301–314.
26. H. Valero, E. Biocca, and L. Cocco, *Yanomama: The Story of Helena Valero, a Girl Kidnapped by Amazonian Indians* (New York: Kodansha, 1997).
27. This terminology comes from H. H. Turney-High, *Primitive War: Its Practice and Concepts* (Columbia, SC: University of South Carolina Press, 1971). This pioneering study, originally published in 1949, was the first systematic study of primitive warfare in the anthropological literature.

28. Plato *Phaedrus*, in J. M. Cooper and D. S. Hutchinson, eds., *Complete Works* (Indianapolis: Hackett, 1997).
29. R. W. Wrangham, "Is military incompetence adaptive?" *Evolution and Human Behavior* 20 1999: 3–17.
30. This definition is a modified version of one suggested by R. Brian Ferguson, professor of anthropology at Rutgers University. According to Ferguson, war is "organized, purposeful group action, directed against another group that may or may not be organized for similar action, involving the actual or potential application of lethal force." R. B. Ferguson, "Introduction: Studying War," in *Warfare, Culture, and Environment* (Orlando: Academic Press, 1984), 5. I am also indebted to Van der Dennen's analysis in J. Van der Dennen, *The Origin of War: The Evolution of a Male-Coalitional Reproductive Strategy* (Westport, CT: Greenwood Press, 1996).
31. J. Keegan, *War and Our World* (New York: Vintage, 1998), 72.
32. Title 22, US Code, Section 2656f(d).
33. Z. Brzezinski, *Out of Control: Global Turmoil on the Eve of the Twenty-first Century* (New York: Scribner, 1993). C. Hedges, *What Every Person Should Know About War* (New York: Free Press, 2003).
34. H. Fischer, "Collateral damage," in R. Gutman, and D. Rieff, *Crimes of War: What the Public Should Know* (New York: Norton, 1999).
35. C. Hedges, *What Every Person Should Know About War*, 99–100.
36. Ibid., 100.
37. P. W. Singer, *Children at War* (New York: Pantheon, 2005).
38. H. K. Ullman, *Shock and Awe: Achieving Rapid Dominance* (Washington, DC: NDU Press, 1996), 31.
39. Rome statute of the international criminal court, article 7. The United States, Israel, and China have exempted themselves from the jurisdiction of the International Criminal Court.
40. Psalms 137:9, New Revised Standard Version. C. Chinken, "Rape and sexual abuse of women in international law," *European Journal of International Law* 5 no. 3: 1–17. "Rape victims' babies pay the price of war," *Guardian*, April 16, 2000.
41. R. Lemkin, *Axis Rule in Occupied Europe: Laws of Occupation—Analysis of Government-Proposals for Redress* (Washington, DC: Carnegie Endowment for International Peace, 1944), 79.
42. The Convention on the Prevention and Punishment of the Crime of Genocide was adopted by Resolution 260 (III) A of the U.N. General Assembly on December 9, 1948. *United Nations Treaty Series*, o. 1021, vol. 78 (1951), 277.
43. R. J. Rummel, *Death by Government* (New York: Transaction, 1997). Z. Brzezinski, *Out of Control.*

44. J. A. Black, G. Cunningham, E. Robson, and G. Zólyomi, *The Electronic Corpus of Sumerian Literature*, http://www-etcsl.orient.ox.ac.uk/.
45. G. C. Conroy, *Reconstructing Human Origins*, 2nd ed. (New York: Norton, 2005). In 1953 the celebrated Piltdown Man was exposed as a hoax. The bone fragments turned out to be a motley collection of two medieval human skulls, an elephant molar, a hippopotamus tooth, an orangutan jaw, and a chimpanzee tooth! For the story of Piltdown Man, see R. Millar, *The Piltdown Men* (New York: St. Martin's Press, 1972).
46. R. A. Dart, "The predatory transition from ape to man," *International Anthropological and Linguistic Review* 1 (1953): 207–208.
47. C. K. Brain, *The Hunters or the Hunted? An Introduction to African Cave Taphonomy* (Chicago: University of Chicago Press, 1981); E. S. Vrba, "The significance of bovid remains as indicators of environment and predation patterns," in A. K. Behrensmeyer and A. P. Hill, eds., *Fossils in the Making: Vertebrate Taphonomy and Paleoecology* (Chicago: University of Chicago Press, 1980). L. R. Berger and R. J. Clarke, "The load of the Taung child," *Nature* 379 no. 29 (1996), 778.
48. D. Hart and R. W. Sussman, *Man the Hunted: Primates, Predators and Human Evolution* (New York: Westview, 2005), xv.
49. B. Ehrenreich, *Blood Rites: The Origins and History of the Passions of War* (New York: Henry Holt, 1997), 120.
50. Hart and Sussman, *Man the Hunted*, xviii.
51. This phrase is from D. C. Dennett, *Elbow Room: The Varieties of Free Will Worth Wanting* (Cambridge, MA: MIT Press, 1984).
52. Hart and Sussman, *Man the Hunted*, xviii.
53. This recommendation was initially presented in chapter 4 of Bryan's book *In His Image* (New York: Fleming H. Revell, 1922) and reiterated in his pamphlet *The Menace of Darwinism* (New York: Fleming H. Revell, 1922).
54. W. J. Bryan, *Bryan's Last Speech: Undelivered Speech to the Jury in the Scopes Trial* (Oklahoma City: Sunlight Publishing Society, 1925).
55. C. W. Dugger, "Religious riots loom over Indian politics," *New York Times*, July 27, 2002.

2. EINSTEIN'S QUESTION

1. From Milne's 1934 pamphlet *Peace with Honor*, cited in L. LeShan, *The Psychology of War* (New York: Helios, 1992), 160.
2. A. Einstein, *Ideas and Opinions* (New York: Bonanza Books, 1955), 110.

3. O. Nathan and H. Norden, *Einstein on Peace* (New York: Schocken Books, 1960), 186.
4. A. Einstein and S. Freud, *Why War?* in *The Complete Psychological Works of Sigmund Freud,* vol. 12. (London: Hogarth Press and the Institute of Psycho-Analysis, 1964), 199–201.
5. R. W. Clark, *Freud: The Man and the Cause* (London: Jonathan Cape and Weidenfield & Nicholson, 1980), 486.
6. Ibid., 489. Freud may have had in mind Heine's couplet from the poem "Almansor": "Das war ein Vorspiel nur, dort wo man Bücher/Verbrennt, verbrennt man auch am Ende Menschen." ("That was only a prelude, there where books are burnt, ultimately people are burnt.") A Muslim speaks it while Catholics are burning a copy of the Koran.
7. Einstein and Freud, *Why War?* 213.
8. P. D. Nolan, "Toward an ecological-evolutionary theory of the incidence of warfare in preindustrial societies," *Sociological Theory* 2 no. 1 (2003): 18–30.
9. L. Keeley, *War Before Civilization: The Myth of the Peaceful Savage* (Oxford: Oxford University Press, 1996), 18.
10. S. A. LeBlanc, *Constant Battles: The Myth of the Peaceful, Noble Savage* (New York: St. Martin's Press, 2003), 3. See also C. R. Ember, "Myths about hunter-gatherers," *Ethnology* 17 (1978): 439–448. For a survey of writings on the psychology of war and violence see J. Van der Dennen, *The Origin of War: The Evolution of a Male-Coalitional Reproductive Strategy* (Westport CT: Greenwood Press, 1996). R. A. Gabriel, *No More Heroes: Madness and Psychiatry in War* (New York: Hill and Wang, 1987).
11. D. Hume, *Treatise of Human Nature* (London: Penguin, 1985).
12. D. Woodridge, *The Machinary of the Brain* (New York: McGraw Hill, 1963); D. R. Hofstadter, "Can creativity be mechanized?" *Scientific American* 247 (1982): 20–29; and D. C. Dennett, *Elbow Room: The Varieties of Free Will Worth Wanting* (Cambridge, MA: MIT Press, 1996).
13. L. Wu et al., "Recognition of host immune activation by Pseudomonas aeruginosa," *Science* 309 (5735) (July 29, 2005): 774–77.

3. OUR OWN WORST ENEMY

1. *Hamlet*, 2. 2. 115–117.
2. T. Surovell, N. Waguespack, and P. J. Brantingham, "Global archaeological evidence for proboscidean overkill," *Proceedings of the National Academy of Sciences* 102(17) (2005): 6231–6236.

3. For human microenvironments, see D. Pilbeam, "What makes us human?" in S. Jones, R. Martin, et al., eds., *The Cambridge Encyclopedia of Human Evolution* (Cambridge: Cambridge University Press, 1992). For predators specialized for hunting hominid prey, see C. K. Brain, *The Hunters or the Hunted? An Introduction to African Cave Taphonomy* (Chicago: University of Chicago Press, 1981).
4. C. Hedges, *War Is a Force That Gives Us Meaning* (Anchor, 2002); Z. Brzezinski, *Out of Control: Global Turmoil on the Eve of the Twenty-first Century* (New York: Touchstone, 1993); J. Glover, *Humanity: A Moral History of the 20th Century* (New Haven, CT: Yale University Press, 2001).
5. Rousseau, J.-J. "Discourse on the origin of inequality." *The Social Contract and Discourses*. Trans. G. D. H. Cole. (North Clavendon, VT: Charles Tottle, 1993) p. 79. The term "noble savage" originated neither in Rousseau's *Discourse* nor in John Dryden's 1672 "Conquest of Grenada." It first appeared in Marc Lescarbot's *Histoire de la Nouvelle France* (1609).
6. J. Guilane and J. Zammit, *The Origins of War: Violence in Prehistory* (Malden, MA: Blackwell, 2005).
7. T. R. Pickering, T. D. White, and N. Toth, "Brief communication: Cutmarks on a Plio-Pleistocene hominid from Sterkfontein, South Africa," *American Journal of Physical Anthropology* 111 (1989): 579–584. T. D. White, "Cut marks on the Bodo cranium: A case of prehistoric defleshing," *American Journal of Physical Anthropology* 69: 503–310. Y. Fernandez-Jalvo, J. C. Diez, I. Caceres, and J. Rosell, "Human cannibalism in the early Pleistocene of Europe (Gran Dolina, Sierra de Atapuerca, Burgos, Spain)," *Journal of Human Evolution* 37 (1999): 591–622. F. Weidenreich, *The Skull of Sinanthropus Pekinensis: A Comparative Study on a Primitive Hominid Skull* (Pehpei, Chungking: Geological Survey of China, 1943). M. H. Walpoff, *Human Evolution* (New York: McGraw Hill, 1996). Genetic evidence suggests that cannibalism was probably very widespread. Cannibalism is a risky practice. In the late nineteenth century a New Guinea tribe called the Fore were decimated by a degenerative disease called kuru, similar to mad cow disease, which was caused the practice of ritual consumption of the brains of their deceased. Nearly a century later, researchers at the University of London published the results of there genetic analysis of several elderly Fore women who survived the kuru epidemic. They found that all three women possessed a particular genetic signature that gave them immunity from the disease. This genetic signature is very widespread in human populations and appears to have evolved specifically as a protection against the ill effects of cannibalism, enabling our ancestors to devour their neighbors with impunity. S. Mead, et al., "Balancing selection at the prion protein gene consistent with prehistoric kuru-

like epidemics," *Science* 300 (5619) (2003): 640–643. However, more recently this research has been sharply criticized on methodological grounds by Marta Soldevila and her colleagues at Pompeu Fabra University in Barcelona, Spain. M. Soldevila, et al., "The prion protein gene in humans revisited: Lessons from a worldwide resequencing study," *Genome Research*, Epub 20 December 2005.

8. E. Trinkaus, "Hard times among the Neanderthals," *Natural History* 87(10) (Dec. 1978): 58–63. L. Bachechi, et al., "An arrow-caused lesion in a Late Upper Palaeolithic human pelvis," *Current Anthropology* 38 (1997): 134–40. F. Wendorf and R. Schild, *The Wadi Kubbaniya Skeleton: A Late Paleolithic Burial from Southern Egypt* (Dallas, TX: Southern Methodist University Press, 1986). F. Wendorf, "Site 117: A Nubian Final Paleolithic graveyard near Jebel Sahaba, Sudan," in *The Prehistory of Nubia*, ed. F. Wendorf, vol. 2, (Dallas, TX: Southern Methodist University Press, 1968).
9. L. H. Keeley, *War Before Civilization* (Oxford: Oxford University Press, 1996). K. R. Otterbein, *How War Began* (College Station, TX: Texas A&M University Press, 2004). R. A. Marlar et al., "Biochemical evidence of cannibalism at a prehistoric Puebloan site in southwestern Colorado," *Nature* 407 (2000): 74–78.
10. For a much more detailed account that is, nonetheless, accessible to the nonspecialist reader, see Guilane and Zammit, *Origins of War.*
11. O. Bar-Yosef, "The Natufian culture in the Levant, threshold to the origins of agriculture," *Evolutionary Anthropology* 6 (1998): 159–177. J. Diamond, *Guns, Germs, and Steel: The Fates of Human Societies* (New York: Norton, 1999). K. F. Otterbein, *How War Began.*
12. R. L. O'Connell, *Ride of the Second Horseman: The Birth and Death of War* (Oxford: Oxford University Press, 1995).
13. J. Keegan, *A History of Warfare* (New York: Vintage, 1993), 161.
14. O'Connell, *Ride of the Second Horseman*, 59. R. J. Wenke, *Patterns in Prehistory: Humankind's First Three Million Years*, 4th ed. (New York: Oxford University Press, 1999).
15. O'Connell, *Ride of the Second Horseman.*
16. H. Breuil and R. Lantier, *The Men of the Old Stone Age* (Westport, CT: Greenwood, 1965). The composite bow, constructed from two varieties of wood, was an immensely important advance in weapons technology. It replaced the simple bow, which was invented around ten thousand years earlier.
17. Keeley, *War Before Civilization.*
18. H. Saggs, *The Might That Was Assyria* (London: Sidgewick & Jackson, 1984), 258.
19. "Lament for Urim," in J. A. Black, G. Cunningham, E. Robson, and G. Zólyomi, *The Electronic Corpus of Sumerian Literature.* http://www-etcsl.orient.ox.ac.uk/.

20. Homer *The Iliad*, trans. R. Feagles (London: Penguin, 1999), Book 20.
21. Herodotus, *The Histories*, trans. Aubrey de Selincourt (London: Penguin, 2003). Guilane and Zammit, *Origins of War.* Some historians believe Herodotus massaged the figures of the number of Persian casualties. The next most deadly battle in the Greco-Roman world took place at Cannae, in Italy, in 216 B.C., where Hannibal's soldiers wiped out 50,000 Roman troops in one day.
22. Exodus 23:27; Deuteronomy 20:16-20.
23. Joshua 10:40.
24. A. Hitler, *Mein Kampf,* trans. Ralph Manheim (London: Hutchinson, 1969). N. H. Baynes, ed., *The Speeches of Adolf Hitler, April 1922–August 1939* (Oxford: Oxford University Press, 1942).
25. For example, Saint Augustine, Saint Gregory, Saint Bernard of Clairvaux, Saint John Chrisostom, and Saint Ambrose. For Luther's anti-Semitic invective, see H. Haile, *Luther* (Garden City, NY: Doubleday, 1980).
26. Matthew 10:34.
27. C. H. Robinson, *The Conversion of Europe* (London: Longman's, Green, 1917).
28. R. Fletcher, *The Barbarian Conversion: From Paganism to Christianity* (Berkeley, CA: University of California Press).
29. J. P. Strayer, "The Albigensian crusades," in F. Chalk and K. Jonassohn, eds., *The History and Sociology of Genocide: Analyses and Case Studies* (Newhaven, CT: Yale University Press, 1990), 119. W. L. Wakefield, *Heresy, Crusade and Inquisition in Southern France, 1100–1250* (London: Allen and Unwin, 1974).
30. D. Crouzet, *Les Guerriers de Dieu: La violence au temps des troubles de religion vers 1525–vers 1610* (Paris: Champvallon, 1990).
31. J. Spence, *God's Chinese Son* (New York: Norton, 1996).
32. Koran 8:65.
33. *Sahih Muslim*, No. 4370. (Lahore: Sh. Mohammad Ashraf, 1978).
34. J. Waller, *Becoming Evil: How Ordinary People Commit Genocide and Mass Killing* (Oxford: Oxford University Press, 2002).
35. Ibid.
36. F. Wertham, *A Sign for Cain: An Exploration of Human Violence* (New York: Macmillan, 1966). M. Coe, *Atlas of Ancient America* (New York: Facts on File, 1986). M. Livi-Bacci, *Concise History of World Population* (London: Blackwell, 1996).
37. Testimony of Robert Brent, Lieutenant James Connor, and John Smith, cited in Waller, *Becoming Evil*, 25–26.
38. Keeley, *War Before Civilization*, 39.
39. D. P. Fry, *The Human Potential for Peace: An Anthropological Challenge to Assumptions About War and Violence* (Oxford: Oxford University Press, 2006).
40. This study is sharply criticized in Fry, *The Human Potential for Peace.*

41. Keeley, *War Before Civilization.*
42. S. A. LeBlanc, *Constant Battles: The Myth of the Peaceful, Noble Savage* (New York: St. Martin's Press, 2003), 155.
43. M. J. Leahy, *Exploration into Highland New Guinea, 1930–1935* (Tuscaloosa, AL: University of Alabama Press). J. Ross, "Effects of contact on revenge hostilities amongst the Achuara Jivaro," in R. Ferguson, ed., *Warfare, Culture and Environment* (Orlando, FL: Academic Press, 1984). N. Chagnon, *Studying the Yanomamo* (New York: Holt, Rinehart & Winston, 1974). M. Meggitt, *Blood Is Their Argument* (Palo Alto, CA: Mayfield, 1977). K. Heider, *The Dugum Dani* (Chicago: Aldine de Guyter, 1970). These and many other pertinent statistics can be found in LeBlanc, *Constant Battles.*
44. R. K. Dentan, *The Semai: A Nonviolent People of Malaya* (New York: Holt, Rinehart & Winston, 1968), 58–59. This passage has often be quoted irresponsibly by authors with an ax to grind. For an excellent discussion, see C. A. Robarchek and R. K. Dentan, "Blood drunkenness and the bloodthirsty Semai: Unmasking another anthropological myth," *American Anthropologist* 98 no. 2 (1987): 356–365.

4. THE ORIGINS OF HUMAN NATURE

1. S. Freud, *Civilization and Its Discontents,* in *The Standard Edition of the Complete Psychological Works of Sigmund Freud,* vol. 21 (London: Hogarth Press, 1961).
2. S. Freud, *Beyond the Pleasure Principle,* in *The Standard Edition of the Complete Psychological Works of Sigmund Freud,* vol. 18 (London: Hogarth Press, 1955), 197.
3. M. R. Davie, *The Evolution of War: A Study of its Role in Early Societies* (Port Washington, NY: Kennikat Press, 1929), 9.
4. Davie cites Malthus, correctly understands war as a tool in the struggle for survival, and makes extensive use of the writings of the evolutionary sociologist William Graham Sumner, but he seems to studiously avoid using Darwinian thinking to explain the rich anthropological data that he presents.
5. T. Dobzhansky, "Nothing in biology makes sense except in the light of evolution," *American Biology Teacher* 35 (March 1973): 125–129.
6. W. Paley, *Natural Theology; or, Evidences of the Existence and Attributes of the Deity,* 12th ed. (London: J. Faulder, 1809), 413.
7. T. R. Malthus, *An Essay on the Principle of Population,* 1st ed., ed. A. Flew (Harmondsworth: Penguin Books, 1798 [1970], 71).

8. C. Darwin, *The Autobiography of Charles Darwin* (New York: Norton, 1958), 120.
9. C. Darwin, *On the Origin of Species by Means of Natural Selection* (Cambridge, MA: Harvard University Press, 2005).
10. C. Darwin, *The Correspondence of Charles Darwin*, vol. 8 (Cambridge: Cambridge University Press, 1993), 224. The digger wasp, described in chapter 2, is a member of this diverse family of insects.
11. C. Zimmer, *Parasite Rex* (New York: Touchstone, 2000), xv–xvi.
12. A. Schopenhauer, "On the vanity and suffering of life," in R. Taylor, ed., *The Will to Live: Selected Writings of Arthur Schopenhauer* (New York: Continuum, 1990), 208.
13. Actually, there are other components of evolutionary change such as genetic drift, founder effects, and so on. I refrain from discussing these, not because they are unimportant, but because they are not directly relevant to the phenomenon of adaptation.
14. M. E. P. Seligman, "Phobias and preparedness," *Behavior Therapy* 2 (1971): 307–320. The idea is not that we have an instinctive fear of these creatures, but rather than we learn to be frightened of them more easily than we learn to be frightened of other items that did not pose a threat during the Stone Age. The same learning bias has been observed in nonhuman primates.
15. K. Lorenz, *On Aggression* (New York: Bantam, 1966). H. Markl, "Aggression und Beuteverhalten bei Piranhas (Sera-salminae)," *Zeitschrift für Tierpsychologie* 30 (1972): 190–216. L. R. Baxter et al., "Brain mediation of Anolis social dominance displays," *Brain, Behavior and Evolution* 57(4) (2001): 169–183.
16. There is a large literature on nonhuman infanticide. For an excellent review, see G. S. Hausfater and S. B. Hrdy, eds., *Infanticide: Comparative and Evolutionary Perspectives* (New York: Aldine, 1984). For nonhuman predation, see G. A. Polis, "The evolution and dynamics of intraspecific predation," *Annual Review of Ecology Systematics* 12 (1981): 225–251. For the "murder rate" amongst nonhuman species, see G. C. Williams, "Huxley's evolution and ethics in sociobiological perspective," *Zygon* 23 (1988): 383–407; S. J. Gould, "A thousand acts of kindness," in *Eight Little Piggies* (New York: Norton, 1993).
17. R. Wrangham, "Evolution of coalitionary killing," *Yearbook of Physical Anthropology* 42 (1999): 1–30. J. A. Byers, *American Pronghorn: Social Adaptations and the Ghosts of Predators Past* (Chicago: University of Chicago Press, 1997). T. H. Clutton-Brock, F. E. Guinness, and S. D. Albon, *Red Deer: Behavior and Ecology of Two Sexes* (Chicago: University of Chicago Press, 1982).
18. H. Topoff et al., "Behavioral adaptations for raiding by the slave-making ant *Polyergus breviceps*," *Journal of Insect Behavior* 2 (1989): 545–556. R. Wrangham,

"Evolution of coalitionary killing," *Yearbook of Physical Anthropology* 42 (1999): 1–30. I. D. Mech et al., *The Wolves of Denali* (Minneapolis, MN: University of Minnesota Press, 1998). C. Packer et al., "Reproductive success of lions," in T. H. Clutton-Brock, ed., *Reproductive Success: Studies of Individual Variation in Contrasting Breeding Systems* (Chicago: University of Chicago Press, 1988). H. Kruuk, *The Spotted Hyena: A Study of Predation and Social Behavior* (Chicago: University of Chicago Press, 1972). T. M. Caro and D. A. Collins, "Male cheetahs of the Serengetti," *National Geographic Research* 2 (1986): 75–86.

19. E. O. Wilson, *Sociobiology: The New Synthesis*, 25th anniversary edition (Cambridge, MA: Harvard University Press, 2000), 249.
20. M. N. Muller and R. W. Wrangham, "Dominance, cortisol and stress in wild chimpanzees (*Pan troglodytes schweinfurthii*)," *Behavioral Ecology and Sociobiology* 55 (2004): 332–340.
21. The terms "baboon pattern" and "chimpanzee pattern" are from J. Van der Dennen, *The Origin of War: The Evolution of a Male-Coalitional Reproductive Strategy* (Westport CT: Greenwood Press, 1996). The account of battling olive baboons is from J. Van Hooff, "Intergroup competition and conflict in animals and man," in J. Van der Dennen and V. Falger, eds., *Sociobiology and Conflict: Evolutionary Perspectives on Competition, Cooperation, Violence and Warfare* (London: Chapman and Hall, 1990), 38–39.
22. J. C. Mitani and D. P. Watts, "Correlates of territorial boundary-patrol behavior in wild chimpanzees," *Animal Behavior* 70 (2005): 1079–1086.
23. S. O'Connell, "Apes of war . . . is it in our genes?" *Telegraph*, July 1, 2004.
24. R. W. Wrangham and D. Peterson, *Demonic Males: Apes and the Origins of Human Violence* (Boston: Houghton-Mifflin, 1996).
25. J. Itani, "Intraspecific killing among nonhuman primates," *Journal of Social and Biological Structure* 5 (1982): 361–368.
26. R. W. Wrangham, "African apes: The significance of African apes for reconstructing social evolution," in W. G. Kinzey, ed., *The Evolution of Human Behavior: Primate Models* (Albany, NY: SUNY Press, 1987).
27. Although there are many genuine distinctions between chimpanzee and bonobo behavior, these differences are often exaggerated, by scientists and lay people alike. For a good survey, see C. B. Stanford, "The social behavior of chimpanzees and bonobos: Empirical evidence and shifting assumptions," *Current Anthropology* 19 no. 4 (1998): 399–420.
28. Wrangham and Peterson, *Demonic Males*, 204. See also C. B. Stanford, *The Hunting Apes: Meat Eating and the Origins of Human Behavior* (Princeton, NJ: Princeton University Press, 1999).

29. For ideas of resources, see P. Meyer, "Ethnocentrism in human social behavior: Some biosociological considerations," in V. Reynolds, V.S.E. Falger, and I. Vine, eds., *The Sociobiology of Ethnocentrism: Evolutionary Dimensions of Xenophobia, Discrimination, Racism and Nationalism* (London: Croom Helm, 1987).
30. Quoted in Isabel Lyon's journal, February 3, 1906. Lyon was Twain's secretary. See K. Lystra, *Dangerous Intimacy: The Untold Story of Mark Twain's Final Years* (Berkeley: University of California Press, 2004).
31. R. W. Wrangham, "Is military incompetence adaptive?" *Evolution and Human Behavior* 20 (1999): 5.
32. Ibid., 6.
33. V. Woolf, *The Three Guineas* (New York: Harvest, 1963), 13.
34. J. S. Goldstein, *War and Gender: How Gender Shapes the War System and Vice Versa* (Cambridge: Cambridge University Press, 2001). E. Hancock, "Women as killers and killing women: The implications of 'gender-neutral' armed forces," in M. Evans and A. Ryan, eds., *The Human Face of Warfare* (St. Leonards, NSW: Allen & Unwin, 2002).
35. L. Cottrell, *The Warrior Pharaohs* (London: Evans Brothers, 1968), 52.
36. R. Pearsall, *Night's Black Angels: The Forms and Faces of Victorian Cruelty* (London: Hodder & Stoughton, 1975).
37. E. O. Wilson, *Sociobiology*.
38. D. M. Buss, *The Murderer Next Door: Why the Mind Is Designed to Kill* (New York: Penguin, 2005). Hilton N. Z. et al., "The functions of aggression by male teenagers," *Journal of Personality and Social Psychology* 79 (2000): 988–994. J. Hyde, "Gender differences in aggression," in J. S. Hyde and M. C. Linn, eds., *The Psychology of Gender: Advances Through Meta-Analysis* (Baltimore: Johns Hopkins, 1986); D. F. Bjorkqvist et al., "Do girls manipulate and boys fight? Developmental trends in regard to direct and indirect aggression," *Aggressive Behavior* 18 (1992): 117–127.
39. T. W. Smith, "The polls: Gender and attitudes towards violence," *Public Opinion Quarterly* 48 (1984): 384–396. J. S. Goldstein, *War and Gender.*
40. J. W. Dower, *War Without Mercy: Race and Power in the Pacific War* (New York: Pantheon, 1986).
41. J. Bourke, *An Intimate History of Killing: Face to Face Killing in 20th-Century Warfare* (New York: Basic Books, 1999), 149.
42. Ibid., 300. Swanwick's original comment appeared in her book *I Have Been Young* (London: Victor Gollancz, 1935).
43. J. P. Gray, "Ethnographic atlas codebook," 1998 *World Cultures* 10, no. 1 (1998): 86–136.

44. L. L. Betzig, *Despotism and Differential Reproduction: A Darwinian View of History* (New York: Aldine, 1986).
45. H. M. Cooper, A. Munich, and S. Squier, eds., *Arms and the Woman: War, Gender and Literary Representation* (Chapel Hill, NC: University of North Carolina Press, 1986). J. S. Goldstein, *War and Gender.*
46. R. L. O'Connell, *Of Arms and Men: A History of War Weapons and Aggression* (Oxford: Oxford University Press, 1989), 152.
47. C. Hedges, *War Is a Force That Gives Us Meaning* (New York: Anchor, 2002), 100–102.
48. Cited in R. Holmes, *Acts of War: The Behavior of Men in Battle* (London: Free Press, 1985), 93.
49. Holmes, *Acts of War*.
50. Numbers 31:16–18.
51. Tacitus *The Histories* (Harmondsworth: Penguin, 1982), 156.
52. W. Rodzinski, *A History of China* (Oxford: Pergamon, 1979), 164–165.
53. It was partly in an effort to stem the tide of rape that the French pursued a policy of state-approved brothels (*Bordels Militaires de Campagne*, or BMCs). Two prostitutes were recommended for the Croix de Guerre—a decoration awarded for acts of heroism in battle—for making a thirty-mile trek in forty-eight hours to relieve a remote military outpost. R. Holmes, *Acts of War*.
54. A. Beevor, "They raped every woman from eight to 80," *Guardian*, May 1, 2002. R. Gellately and B. Kiernan, "The study of mass murder and genocide," in R. Gellately and B. Kiernan, eds., *The Specter of Genocide: Mass Murder in Historical Perspective* (Cambridge: Cambridge University Press, 2003). P. Schrijueus, *The Crash of Ruin: American Combat Soldiers in Europe During World War* II (New York: New York University Press, 1998).
55. G. Rodrique, "Sexual violence: enslavement and enforced prostitution," in R. Gutman and D. Rieff, *Crimes of War: What the Public Should Know* (New York: Norton, 1999). N. Pope, "Torture," in ibid.
56. J. Hatzfeld, *Machete Season: The Killers in Rwanda Speak* (New York: Farrar, Straus & Giroux, 2005), 134.
57. I. Chang, "The rape of Nanking," in A. L. Barstow, ed., *War's Dirty Secret: Rape, Prostitution and Other Crimes Against Women* (Cleveland, OH: Pilgrim Press, 2000), 47.
58. L. Keeley, *War Before Civilization: The Myth of the Peaceful Savage* (Oxford: Oxford University Press, 1996), 86.

5. HAMLET'S QUESTION

1. *Hamlet*, 2. 2. 115–117.
2. Genesis, 3:19.
3. Plato *Phaedo*, 105c.
4. C. McGinn, *The Mysterious Flame* (New York: Basic Books, 2000).
5. P. Churchland, A *Neurocomputational Perspective* (Cambridge, MA: Bradford/MIT, 1984).
6. I am using the word "picture" here to do duty for any form of neural representation. I do not wish to imply that neural representations are literally pictures in the brain.
7. P. Boyer, *Religion Explained: The Evolutionary Origins of Religious Thought* (New York: Basic Books, 2001), 102.
8. Many of the details of the modular theory of mind remain controversial and obscure. We do not know exactly how many of these modules populate our brains. There may be hundreds, or even thousands of them. We do not know if the mind is only partially modular, or whether it is entirely ("massively") modular. Thankfully, for the purposes of this discussion I can ignore most of these arcane debates and focus on the big picture. For an excellent summary of the main controversies, see R. Samuels, "Massively modular minds: Evolutionary psychology and cognitive architecture," in P. Carruthers and A. Chamberlain, eds., *Evolution and the Human Mind: Modularity, Language and Meta-Cognition* (Cambridge: Cambridge University Press, 2000).
9. C. McGinn, *Mental Content* (Oxford: Basil Blackwell, 1989). McGinn builds on the pioneering contributions of Kenneth Craik for this approach to human cognition. K. Craik, *The Nature of Explanation* (Cambridge: Cambridge University Press, 1967).
10. The quotation is from G. Leibniz, *New Essays on Human Understanding* (Cambridge: Cambridge University Press, 1996). This was written in 1704 as a blow-by-blow critique of John Locke's *Essay Concerning Human Understanding* but was withheld from publication because of Locke's death; it was published posthumously in 1765.
11. See F. Attneave, "In defense of homunculi," in W. Rosenblith, ed., *Sensory Communication* (Cambridge, MA: MIT Press, 1960); D. C. Dennett, *Brainstorms* (Cambridge, MA: Bradford Books, 1978). J. A. Fodor, "Methodological solipsism considered as a research strategy in cognitive psychology," *Behavioral and Brain Sciences* 3 (1980): 63–73.
12. D. C. Dennett, *Freedom Evolves* (London: Penguin, 2004), 162.

13. D. Hume, *Natural History of Religion* (Palo Alto, CA: Stanford University Press, 1957), 30.
14. P. S. Churchland, *Brainwise: Studies in Neurophilosophy* (Cambridge, MA: Bradford/MIT, 2002), 40–41.
15. G. Leibniz, "The monadology," in *The Monadology and Other Philosophical Writings*, trans. R. Latta (Oxford: Oxford University Press, 1898), paragraph 17.
16. I. Kant, *Critique of Pure Reason* (Cambridge: Cambridge University Press, 1999). The University of London psychologist Henry Plotkin has fruitfully explored the connection between evolutionary psychology and Kantian thought in *Darwin Machines and the Nature of Knowledge* (Cambridge, MA: Harvard University Press, 1997).

6. A LEGACY OF LIES

1. The story was written in 1905, as a response to the American intervention in the Philippines, but was rejected by *Harper's Bazaar*. It was posthumously published in 1916, during World War I.
2. M. Twain, "The War Prayer," in *Collected Tales, Sketches, Speeches and Essays, 1891–1910* (New York: Library of America, 1992), 654–655. Ironically, Twain's friends and family begged him not to publish "The War Prayer." He nonetheless submitted it to *Harper's Bazaar* in 1905, and it was rejected. "Only the dead," Twain bitterly remarked, "are permitted to tell the truth." A. B. Paine, *Mark Twain, a Biography* (New York: Harper & Brothers, 1912), 54.
3. A. Huxley, *The Olive Tree* (New York: Harper, 1937), 85–86.
4. H. A. Bosmajian, *The Language of Oppression* (New York: University Press of America, 1983). W. C. Gay, "The language of war and peace," in L. Kurtz, ed., *Encyclopedia of Violence, Peace, and Conflict*, vol. 2 (San Diego: Academic Press, 1999). A. Bandura, "Moral disengagement in the preparation of inhumanities," *Personality and Social Psychology Review* 3 (1999): 193–209. W. Lutz, *Doublespeak: From Revenue Enhancement to Terminal Living: How Government, Business, Advertisers, and Others Use Language to Deceive You* (New York: HarperCollins, 1990). N. Werth, *The Mechanism of a Mass Crime: The Great Terror in the Soviet Union, 1937–1938*, in R. Gellately and B. Kiernan, eds., *The Specter of Genocide: Mass Murder in Historical Perspective* (Cambridge: Cambridge University Press, 2003). J. Coughlin and C. Kuhlman, *Shooter: The Autobiography of the Top-Ranked Marine Sniper* (New York: St. Martin's Press, 2005). D. Baum, "The price of valor," *New Yorker*, July 12 and 19, 2004, 3.

5. Plato discusses "know thyself" in *Charmides,* 164d–165b. For Diogenes Laertius's remarks on Thales, see D. Lacrtius, *Lives of the Philosophers* (Washington, DC: Regnery Publishing Inc., 1969).
6. H. Fingarette, *Self-Deception* (London: Routledge & Kegan Paul, 1969), 1. See also D. L. Smith, *Why We Lie: The Evolutionary Roots of Deception and the Unconscious Mind* (New York: St. Martin's Press, 2004).
7. These remarks are found in vol. 25:252 and vol. 25:863, cf. vol.: 477–478, 865 in *Kant's gesammelte Schriften* (Berlin: Georg Reimer, 1900–1942).
8. J. W. von Goethe, "Significant help given by an ingenious turn of phrase," in *Scientific Studies,* trans. D. Miller (New York: Suhrkamp, 1988), 39, and "Sprichwörtlich," in *Münchner Ausgabe: Sämtliche Werke nach Epochen seines Schaffens,* vol. 9 (München: C. Hanser, 1985), 144. For the literature on depression, self-deception, and mental health see R. D. Lane, K. R. Merikangas, G. E. Schwartz et al., "Inverse relationship between defensiveness and lifetime prevalence of psychiatric disorder," *American Journal of Psychiatry* 147 (1990): 573–578. H. A. Sackheim and R. C. Gur, "Self-deception, other deception, and self-reported psychopathology," *Journal of Consulting Clinical Psychology* 47 (1979): 213–215. H. A. Sackheim and A. Z. Wegner, "Attributional patterns in depression and euthymia," *Archives of General Psychiatry* 43 (1986): 553–560. S. E. Taylor and J. D. Brown, "Illusion and well-being: a social psychological perspective on mental health," *Psychological Bulletin* 103 (1988): 193–210; P. M. Lewinsohn et al., "Social competence and depression: the role of illusionary self-perceptions," *Journal of Abnormal Psychology* 89 (1980): 203–212. L. B. Alloy and L. Y. Abramson, "Depression and pessimism for the future: biased use of statistically relevant information in predictions of self versus others," *Journal of Personality and Social Psychology* 41 (1987): 1129–1140; ibid., "Judgement of contingency in depressed and nondepressed students: Sadder but wiser?" *Journal of Experimental Psychology: General* 108 no. 4 (1979): 441–485. Ibid., "Learned helplessness, depression and the illusion of control," *Journal of Personality and Social Psychology* 42 (1982): 1114–1126.
9. Twain to W. D. Howells, August 31, 1884, in A. B. Paine, *Mark Twain's Letters* (New York: Classic Books, 2000).
10. Demosthenes, "Third Olynthiac," *Olynthiacs, Philippics, Minor Public Speeches,* trans. J. H. Vince (Cambridge, MA: Harvard University Press, 1985), 53.
11. Z. Kunda, "Motivated inference: Self-serving generation and evaluation of causal theories," *Journal of Personality and Social Psychology* 53 (1987): 636–47.
12. D. Dunning, C. Heath, and J. M. Suls, "Flawed self-assessment: Implications for health, education and the workplace," *Psychological Science in the Public Interest* 5 no. 3: 69–106, 69.

13. For surgeons, see D. A. Risucci, A. J. Torolani, and R. A. Ward, "Ratings of surgical residents by self, supervisors and peers," *Surgical Gynecology and Obstetrics* 169 (1989): 519–526. For college students' love affairs, see T. K. MacDonald and M. Ross, "Assessing the accuracy of predictions about dating relationships. How and why do lovers' predictions differ from those of observers?" *Personality and Social Psychology Bulletin* 25 (1999): 1417–1429. For lawyers, see E. F. Loftus and W. A. Wagenaar, "Lawyers' predictions of success," *Jurimetrics Journal* 29 (1988): 437–453. For doctors, see T. Odean, "Volume, volatility, price and profit when all traders are above average," *Journal of Finance* 8 (1998): 1887–1934. For unbiased self-assessment, see J. Friedrich, "On seeing oneself as less self-serving than others: the ultimate self-serving bias?" *Teaching of Psychology* 23 (1996): 107–109; and E. Pronin, D. Y. Linn, and I. Ross, "The bias blind spot: Perceptions of bias in self versus others." *Personality and Social Psychology Bulletin* 28 (2002): 369–381.
14. For general discussions of how mental division makes self-deception possible, see D. Davidson, "Deception and division," in R. Wollheim and J. Hopkins, eds., *Philosophical Essays on Freud* (Cambridge: Cambridge University Press, 1982), and R. Kurzban and C. A. Aktipis, "Modular minds, multiple motives," in M. Shaller, J. Simpson, and D. Kenrick, eds., *Evolution and Social Psychology* (London: Psychology Press, 2005). For an influential account of self-deception that does not invoke mental division or the simultaneous presence of contradictory beliefs, see A. Mele, *Self-Deception Unmasked* (Princeton: Princeton University Press, 2001).
15. S. P. Springer and G. Deutsch, *Left Brain, Right Brain Perspectives from Cognitive Neoroscience* (San Francisco: W. H. Freeman), 1997.
16. Cited in G. Marcus, *The Birth of the Mind: How a Tiny Number of Genes Creates the Complexities of Human Thought* (New York: Basic Books, 2004), 89.
17. M. Gazzaniga, *The Ethical Brain* (Chicago: University of Chicago Press, 2005).
18. J. F. Wilkins and D. Haig, "What good is genomic imprinting: The function of parent-specific gene expression," *Nature Reviews Genetics* 4 (2003): 359–368. D. Haig, "Placental hormones, genomic imprinting, and maternal—fetal communication," *Journal of Evolutionary Biology* 9 (1996): 357–380. Ibid., "Genetic conflicts in human pregnancy," *Quarterly Review of Biology* 68 (1993): 495–532.
19. E. Kverne et al., "Genomic imprinting and the differential roles of parental genomes in brain development," *Developmental Brain Research* 92 (1996): 91. E. Keverne et. al., "Primate brain evolution: Genetic and functional considerations," *Proceedings of the Royal Society London B* 262 (1996): 689.

20. See D. Haig, "On intrapersonal reciprocity," *Evolution and Human Behavior* 24 (2003): 418–425; and A. Burt and R. Trivers, *Genes in Conflict: The Biology of Selfish Genetic Elements* (Cambridge, MA: Belknap, 2006).
21. For evidential difficulties, see A. Grunbaum, *Validation in the Clinical Theory of Psychoanalysis: A Study in the Philosophy of Psychoanalysis* (New York: International Universities Press, 1993), and E. Erwin, *A Final Accounting: Philosophical and Empirical Issues in Freudian Psychology* (Cambridge, MA: MIT, 1995). For the anachronistic character of Freudian theory, see P. Kitcher, *Freud's Dream: A Complete Interdisciplinary Science of Mind* (Cambridge, MA: MIT, 1995). For Freudian dishonesty, see A. Esterson, *Seductive Mirage* (Chicago: Open Court, 1993). For an introductory review of the critical literature, see D. L. Smith, *Psychoanalysis in Focus* (London: Sage, 1999).
22. For a sophisticated analysis of how Libet's results might be interpreted, see D. C. Dennett, *Consciousness Explained* (Boston: Back Bay Books, 1992).
23. For the details, see D. L. Smith, *Freud's Philosophy of the Unconscious* (Dordrecht, NL: Kluwer, 1999), and D. L. Smith, "'Some unimaginable substratum': a Contemporary introduction to Freud's philosophy of mind," in M. C. Chung and C. Feltham, eds., *Psychoanalytic Knowledge and the Nature of Mind* (New York: Palgrave, 2003). There is a substantial scientific literature supporting Freud's hypothesis that consciousness is mediated by inner speech. For an excellent review, see A. Morin, "Possible links between self-awareness and inner speech: Theoretical background, underlying mechanisms and empirical evidence," *Journal of Consciousness Studies* 12 nos. 4–5 (2005): 115–134.
24. Plato *Republic*, trans. T. Griffith (Cambridge: Cambridge University Press, 2000), 343d–e.
25. Ibid., 361a–b.
26. Ibid., 265b–d.
27. T. Hobbes, *Leviathan* (London: Penguin, 1981), 186.
28. Chapter 2 of my book, *Why We Lie*, details many examples of nonhuman deception.
29. M. Twain, *Mark Twain's Autobiography* (New York: Harper Brothers, 1924).
30. R. L. Trivers, "Self-deception in service of deceit," in *Natural Selection and Social Theory: Selected Papers of Robert Trivers* (Oxford: Oxford University Press, 2002).
31. M. R. Banaji and A. G. Greenwald, "Implicit stereotyping and prejudice," in M. P. Zanna and J. M. Olson, eds., *The Psychology of Prejudice: The Ontario Symposium*, vol. 7 (Hillsdale, NJ: Lawrence Erlbaum, 1995). You can take the test yourself at https://implicit.harvard.edu/implicit/.
32. H. E. Adams, L. W. Wright, and B. A. Lohr, "Is homophobia associated with homosexual arousal?" *Journal of Abnormal Psychology* 105 no. 3 (1996): 440–445.

7. MORAL PASSIONS

1. A. Hitler, *Mein Kampf*, trans. Ralph Manheim (London: Hutchinson, 1969), 231–232.
2. L. S. Dawidowicz, *The War Against the Jews: 1933–1945* (New York: Bantam, 1986), 21.
3. F. Ponchaud, *Cambodia: Year Zero* (Harmondsworth: Penguin, 1978), 69.
4. J. Mason, "A brief history of the Pequot war," *Massachusetts Historical Society Collections*, 2nd ser., 8 (Boston: Massachusetts Historical Society, 1826), 140–141.
5. M. Wooldridge, "Can religion be blamed for war?" BBC News, February 24, 2004. "Bush, Saddam turning to God for victory," *Hindustan Times*, April 4, 2003.
6. For Darwin and Twain, see M. Twain, "Our guest," in *Collected Tales, Sketches, Speeches and Essays, 1891–1910* (New York: Library of America, 1992). For Freud and Twain, see P. Gay, *Freud: A Life for Our Time* (New York: Norton, 1998). For James and Twain, see J. G. Horn, *Mark Twain and William James: Crafting a Free Self* (Columbia, MO: University of Missouri Press, 1996).
7. M. Twain, *The Adventures of Huckleberry Finn*, 2nd ed. (New York: Norton, 1967). This wonderful passage was drawn to my attention by David Nyberg's excellent book *The Varnished Truth* (Chicago: University of Chicago Press, 1993).
8. D. Hume, *A Treatise of Human Nature* (London: Penguin, 1985): 520–523.
9. The literature on the famous trolley problem is extensive. See P. Foot, "The problem of abortion and the doctrine of the double effect," in *Virtues and Vices* (Oxford: Basil Blackwell, 1978). S. Kagan, *The Limits of Morality* (Oxford: Oxford University Press, 1989). F. M. Kamm, "Harming some to save others," *Philosophical Studies*, 57 (1989): 227–60. J. J. Thomson, "Killing, letting die, and the trolley problem," *Monist* 59 (1976): 204–217. J. J. Thomson, "The trolley problem," *Yale Law Journal* 94 (1985): 1395–1415. P. Unger, *Living High and Letting Die* (Oxford: Oxford University Press, 1996). For experimental investigation and the automaticity of moral cognition, see M. D. Hauser, *Moral Minds: The Unconscious Voice of Right and Wrong* (New York, Harper Collins, in press); M. Gazzaniga, *The Ethical Brain* (New York: Dana, 2005); and J. Haidt, "The emotional dog and its rational tail: a social intuitionist approach to moral judgment," *Psychological Review* 108 (2001): 814–834.
10. S. Blakeslee, "Cells That Read Minds," *New York Times*, January 10, 2006. M. Iacoboni et al., "Grasping the intentions of others with one's own mirror neuron system," *Public Library of Science, Biology* 3, no. 3., http://biology.plosjournals

.org/. V. Gallese, C. Keysers, and G. Rizzolate, "A unifying view of the basis of social cognition," *Trends in Cognitive Sciences* 8, no. 9 (2004): 337–402.

11. D. Hume, *Treatise of Human Nature* (London: Penguin, 1985), 397. C. Hedges, *War Is a Force That Gives Us Meaning* (New York: Anchor, 2002), 13–14.
12. The injunction to love your neighbor first appears in the nineteenth chapter of the book of Leviticus. In this context, the word "neighbors" was taken literally. It was supposed to refer to members of one's own community. In the Gospel account, Jesus quotes the Old Testament saying but, when asked by an onlooker, "Who is my neighbor?" replies with the parable of the Good Samaritan. The Samaritans were semiconverted pagans who were despised by most Jews. So the Christian version of the saying has greater moral traction than its Old Testament precursor. M. Shermer, *The Science of Good and Evil: Why People Cheat, Gossip, Care, Share and Follow the Golden Rule*. (New York: Henry Holt, 2004).
13. J. Mackie, *Ethics: Inventing Right and Wrong* (London: Penguin, 1999), 83–84.
14. The term "vehicle" when used in this context is from R. Dawkins, *The Selfish Gene* (Oxford: Oxford University Press, 1976).
15. W. D. Hamilton, "The genetical evolution of social behavior: I," *Journal of Theoretical Biology* 7 (1964): 1–16. Ibid., "The genetical evolution of social behavior: II," *Journal of Theoretical Biology* 7 (1964): 17–52. R. L. Trivers, "The evolution of reciprocal altruism," *Quarterly Review of Biology* 46 (1971): 35–57. R. D. Alexander, *The Biology of Moral Systems* (New York: Aldine de Guyter, 1987). The biblical quotation is from Ecclesiastes 11:1.
16. R. Bigelow, *The Dawn Warriors: Man's Evolution Toward Peace* (Boston: Little, Brown, 1969), 3.
17. L. Rodseth et al., "The human community as a primate society," *Current Anthropology* 32, no. 3 (1991): 221–255. M. Shermer, *The Science of Good and Evil*
18. R. Bigelow, *The Dawn Warriors*, 3.
19. Richard D. Alexander first pointed out the uniqueness and significance of human team play. See R. D. Alexander, "Evolution and culture," in N. A. Chagnon and W. G. Irons, eds., *Evolutionary Biology and Human Social Behavior: An Anthropological Perspective* (North Scituate, MA: Duxbury Press, 1979). R. D. Alexander, *Darwinism and Human Affairs* (Seattle: University of Washington Press, 1979). Ibid., *The Biology of Moral Systems*.
20. Joshua 6:21, New International Version.
21. Joshua 7:24–25.
22. For a more technical discussion of these issues, see P. Carruthers, "Distinctively human thinking: Modular precursors and components," in P. Carruthers, S. Lawrence, and S. Stich, eds., *The Innate Mind: Structure and Contents* (Oxford: Oxford University Press, 2005).

23. A. Brinkley, *American History, A Survey*, 9th ed. (New York: McGraw-Hill, 1995), 352.
24. P. Balakian, *Black Dog of Fate: A Memoir* (New York: Basic Books, 1997), quoted in P. Waller, *Becoming Evil: How Ordinary People Commit Genocide and Mass Killing* (Oxford: Oxford University Press, 2002), 53–54.

8. RELUCTANT KILLERS

1. S. Pinker, *The Blank Slate: The Modern Denial of Human Nature* (New York: Viking, 2002). D. M. Buss, *The Murderer Next Door: Why the Mind Is Designed to Kill* (New York: Penguin, 2005).
2. These are the 2005 figures for Jamaica, which has nosed ahead of Colombia and South Africa as the homicide capital of the world.
3. S. L. A. Marshall, *Men Against Fire: The Problem of Battle Command in Future War* (Gloucester, MA: Peter Smith, 1978), 9.
4. Ibid. F. D. G. Williams, *SLAM: The Influence of S. L. A. Marshall on the United States Army* (Fort Monroe, VA: Office of the Command Historian US Army Training and Doctrine Command, 1990). R. Spiller, "S. L. A. Marshall and the ratio of fire," *Journal of the Royal United Services Institute* 133 (1988): 63–71. F. Smoler, "The secret of the soldiers who didn't shoot." *American Heritage* (March 1989), 37–45. H. Leinbaugh, *The Men of Company K* (New York: William Morrow, 1985). R. W. Glenn, *Reading Athena's Dance: Men Against Fire in Vietnam* (Annapolis, MD: Naval Institute Press, 2000). J. Bourke, *An Intimate History of Killing: Face-to-Face Killing in Twentieth-Century Warfare* (New York: Basic Books, 1999). G. Dyer, *War: The Lethal Custom*, rev. ed. (New York: Carroll & Graf, 2004). G. Gurney, *Five Down and Glory* (New York: Putnam's, 1958). For an engaging assessment of Marshall and his work written by his grandson, see Robert Marshall, *Reconciliation Road: A Family Odyssey* (Seattle: University of Washington Press, 2000).
5. For information on abandoned guns, see F. A. Shannon, *The Organization and Administration of the Union Army* (Gloucester, MA: Peter Smith, 1965). The quotation is from G. Dyer, *War*, 55–56. Paddy Griffith interprets the abandoned musket rifles differently, suggesting that they were probably unusable because improperly loaded. See P. Griffith, *Battle Tactics of the Civil War* (New Haven, CT: Yale University Press, 1989).
6. Marshall, *Men Against Fire*, 72.
7. D. A. Grossman, *On Killing* (New York: Little, Brown, 1995), xiv.
8. Williams, *SLAM*. Grossman, *On Killing*.

9. B. Ehrenreich, *Blood Rites: The Origins and History of the Passions of War* (New York: Henry Holt, 1997), 10. They also escape by committing suicide. Almost four hundred Union troops committed suicide during the American Civil War.
10. Ibid.
11. Grossman, *On Killing*. Ibid., "Human factors in war: The psychology and physiology of close combat," in M. Evans and A. Ryan, eds., *The Human Face of Warfare* (St. Leonards, NSW: Allen & Unwin, 2002).
12. P. Orr, *The Road to the Somme: Men of the Ulster Division Tell Their Story* (Belfast: Dufour, 1987), 155. S. Dicks, *From Vietnam to Hell: Interviews with Victims of Post-Traumatic Stress Disorder* (Jefferson, NC: McFarland & Co, 1990), 30.
13. W. Manchester, *Goodbye, Darkness: A Memoir of the Pacific War* (London: Dell, 1980), 17–18.
14. *Henry IV*, 2.3, *Complete Pelican Shakespeare* (London: Penguin, 2002).
15. S.A. Stouffer et. al., *The American Soldier*, vol. 2. (Princeton, NJ: Princeton University Press, 1949). C. Hedges, *What Every Person Should Know About War* (New York: Free Press, 2003). G. Dyer, *War*. E. Wright, *Generation Kill: Devil Dogs, Iceman, Captain America, and the New Face of American War* (New York: Berkley Caliber, 2005). The ancient Egyptian text is cited in R. A. Gabriel, *No More Heroes: Madness and Psychiatry in War* (New York: Hill & Wang, 1987), 47. The account of Epizelus is from book 1 of Herodotus's *The Histories* (Oxford: Oxford University Press, 1998). Plutarch, *Plutarch's Lives: Agesilaus and Pompey Pelopidas and Marcellus* (Cambridge, MA: Harvard University Press, 1917). M. Swanton, ed., *The Anglo-Saxon Chronicle* (New York: Routledge, 1998).
16. Ehrenreich, *Blood Rites*. Holmes, *Acts of War* (London: Free Press, 1985). G. M. Carstairs, "Daru and bhang: Cultural factors in the choice of intoxicant," *Quarterly Journal of Studies on Alcohol* 15 (1954): 220–237. A. Lehmann and L. J. Mihalyi, "Aggression, bravery, endurance and drugs: a radical re-evaluation and analysis of the Masai warrior complex," *Ethnology* 21, no. 4 (1982): 335–347. J. Waller, *Becoming Evil: How Ordinary People Commit Genocide and Mass Killing* (Oxford: Oxford University Press, 2002). R. A. Gabriel, *No More Heroes*. Naval Strike and Air Warfare Center NAVMED P-6410, *Performance Maintenance During Continuous Flight Operations: A Guide for Flight Surgeons* (Fallon, NV: 1 January 2000). C. Hedges, *What Every Person Should Know About War* (New York: Free Press, 2003). A. Beevor, *Berlin: The Downfall, 1945* (London: Penguin, 2002). C. Hedges, *War Is a Force That Gives Us Meaning* (New York: Anchor, 2002). N. Ferguson, *The Pity of War: Explaining World War I* (New York: Basic Books, 1999), 351. R. Graves, *Good-bye to All That: An Autobiography*

(New York: Anchor, 1958), 172. A. Bonadeo, *Mark of the Beast: Death and Degradation in the Literature of the Great War* (Lexington, KY: University Press of Kentucky, 1989), 10. J. Ellis, *The Sharp End: The Fighting Man in World War II* (New York: Charles Scribner's Sons, 1980).

17. E. Baard, "The guilt-free soldier: New science raises the specter of a world without regret," *Village Voice*, January 22, 2003.
18. Gabriel, *No More Heroes*, 73–74. D. A. Grossman, "Human factors in war: The psychology and physiology of close combat." B. P. Dohrenwend et al., "The psychological risks of Vietnam for U.S. veterans: A revisit with new data and methods," *Science* 313 (5789) (2006): 979–982. A. Bonadeo, *Mark of the Beast*, 25. J. E. Mack, *A Prince of Our Disorder: The Life of T. E. Lawrence* (Cambridge, MA: Harvard University Press, 1998). A. W. Lawrence, ed., *T. E. Lawrence by His Friends: A New Selection of Memoirs* (London: McGraw-Hill, 1937), 272.
19. Homer, *The Odyssey*, trans. Robert Fitzgerald (New York: Anchor Doubleday, 1975). Book 12, Lines 125–130.
20. M. Evans, "Close combat: Lessons from the cases of Albert Jacka and Audie Murphy," in M. Evans and A. Ryan, eds., *The Human Face of Warfare*, 47.
21. R. W. Glenn, *Reading Athena's Dance Card*. J. W. Appel and G. W. Beebe, "Preventive psychiatry: an epidemiological approach," *Journal of the American Medical Association* 131 (1946): 1470. R. L. Swank and W. E. Marchand, "Combat neuroses: The development of combat exhaustion," *Archives of Neurology and Psychiatry* 55 (1946): 244. D. A. Grossman, "Human factors in war: The psychology and physiology of close combat." J. Bourke, *An Intimate History of Killing*.
22. D. Baum, "The price of valor," *New Yorker*, July 12 and 19, 2004. D. A. Grossman, "Human factors in war: The psychology and physiology of close combat." J. Atholl, *The Reluctant Hangman: James Berry* (New York: John Long, 1956). R. M. MacNair, *Perpetration-Induced Traumatic Stress: The Psychological Consequences of Killing* (Lincoln, NE: Author's Choice Press, 2005). R. Hoess, *Commandant of Auschwitz: The Autobiography of Rudolf Hoess* (London: Weidenfeld and Nicolson, 1959). R. J. Lifton, *The Nazi Doctors: Medical Killing and the Psychology of Genocide* (New York: Basic Books, 2000). S. Tisdale, "We do abortions here," *Harper's*, October 1987. M. Such-Baer, "Professional staff reaction to abortion work," *Social Casework*, July 1974.
23. B. J. Verkamp, *The Moral Treatment of Returning Warriors in Early Medieval and Modern Times* (Scranton: University of Scranton Press, 1993). S. E. French, *The Code of the Warrior: Exploring Warrior Values, Past and Present* (Lanham, New York, and Oxford: Rowman and Littlefield, 2003). Grossman argues that the high level of ongoing psychiatric casualties in the wake of the Vietnam may

be accounted for by the fact that the unpopularity of the war made it difficult for returning veterans to justify their actions. See Grossman, *On Killing.*

24. Cited in D. Goleman, *Vital Lies, Simple Truths: The Psychology of Self-Deception* (London: Bloomsbury, 1997), 29.
25. P. Caputo, *A Rumor of War* (New York: Henry Holt, 1977), 305–306.
26. I. Kirkpatrick, "Memoirs," cited in J. Bourke, *An Intimate History of Killing,* 208–209.
27. E. Jünger, *The Storm of Steel: From the Diary of a German Storm-Troop Officer on the Western Front* (London: Penguin, 2004), cited in N. Ferguson, *The Pity of War,* 359. T. E. Lawrence, *Seven Pillars of Wisdom: A Triumph* (New York: Anchor, 1991), 511. A. Bonadeo, *Mark of the Beast,* 3–4.
28. J. Bourke, *An Intimate History of Killing,* 16.

9. THE FACE OF WAR

1. Fatal intragroup coalitionary violence does sometimes occur in chimpanzee communities. See K. Fawcett and G. Muhumza, "Death of a wild chimpanzee community member: Possible outcome of intense sexual competition," *American Journal of Primatology* 51 (2000): 243–247; T. Nishida, "The death of Ntologi: The unparalleled leader of M Group," *Pan Africa News* 3 (1996); R. Ktopeni et al., "Ntologi falls??!" *Pan Africa News* 2 (1995): 9–11.
2. J. Goodall, *Through a Window: My Thirty Years with the Chimpanzees of Gombe* (New York: Houghton Mifflin, 1990), 109.
3. Ibid.
4. B. Anderson, *Imagined Communities: Reflections on the Origin and Spread of Nationalism* (New York: Verso, 1983). Katherine Lee Bates, "America the Beautiful." S. V. Mikhalkov and G. G. El-Registan, "Soviet National Anthem." G. R. Johnson, "In the name of the fatherland: an analysis of kin term usage in patriotic speech and literature," *International Political Science Review* 8, no. 2 (1987): 165–174. Koran 49:10.
5. S. Atran, "Strong versus weak adaptationism in cognition and language," in P. Carruthers et. al., eds., *The Innate Mind: Structure and Contents* (Oxford: Oxford University Press, 2005), 145.
6. F. J. Gil-White, "Are ethnic groups biological 'species' to the human brain? Essentialism in our cognition of some social categories," *Current Anthropology* 42, no. 4 (2001): 518.
7. G. Dyer, *War: The Lethal Custom,* rev. ed. (New York: Carroll & Graf, 2004), 128–129.

8. For an excellent review, see H. C. Barratt, "Adaptations to predators and prey," in D. M. Buss, ed., *The Handbook of Evolutionary Psychology* (Hoboken, NJ: Wiley, 2005). In ancient times, the power of agency was distributed far more liberally than it is today—trees, mountains, springs, and rivers were thought to have minds. When it stormed, the sky was angry, and the clouds wept when it rained. And yet, sky, clouds, mountains, and rivers do not behave like agents: they do *not* conform to the distinctive movement pattern of animals. I think that this is best explained along the same lines as Descartes's denial of agency to nonhuman animals. Nobody *spontaneously* treats a mountain as an agent. Belief in thinking mountains and other animistic entities comes about when religious indoctrination overrides evolved common sense. Animism is not the natural condition of mankind: it is every bit as ideologically determined as the belief that Jesus rose from the dead or that Muhammad received the Koran from the angel Gabriel.
9. The term "intentional stance" has gained wide currency in cognitive science and philosophy. The ubiquitous Daniel C. Dennett coined it. See D. C. Dennett, *The Intentional Stance* (Cambridge, MA: MIT, 1989).
10. Philosophically educated readers will notice the doctrine known as "the holism of the mental" lurking behind this discussion. See D. Davidson, *Essays on Actions and Events* (Oxford: Oxford University Press, 2001), and D. C. Dennett, *Brainstorms: Philosophical Essays on Mind and Psychology* (Cambridge, MA: MIT, 1981).
11. S. Mithen, *The Prehistory of the Mind: A Search for the Origins of Art, Religion and Science* (London: Phoenix, 1998), 192. See also M. Douglas, "The pangolin revisited: A new approach to animal symbolism," in R. G. Willis, ed., *Signifying Animals: Human Meaning in the Natural World* (London: Unwin Hyman, 1990), and G. Silberbauer, *Hunter and Habitat in the Central Kalahari Desert* (Cambridge: Cambridge University Press, 1981). It is perhaps worth noting that Mithen may not be correct in calling this "anthropomorphizing." We should not assume that treating nonhuman animals as agents is in any sense an extrapolation from the human case. For empirical support, see S. Atran et al., "Folkbiology doesn't come from folkpsychology: Evidence from Yukatec Maya in cross-cultural perspective," *Journal of Cognition and Culture* 1 (2001): 3–42.
12. Nonhuman animals can also do this to a limited degree, and even reptiles are sensitive to the significance of the direction of a predators' gaze. In most species the program is rigid and sphexish, but in others it is relatively flexible. Dogs provide a remarkable example. Because of their long association with human beings, which goes back at least ten thousand years, domestic dogs have evolved the ability to interpret human beings' mental states and can do this far more

effectively than chimpanzees, who are in most other respects their intellectual superiors. For example, no nonhuman primate is able to "understand" the meaning of a pointing gesture, but dogs—even untrained puppies—grasp it with ease (interestingly, wolves do not have this ability, because there has been no selection pressure on them to acquire it). See J. Krebs and R. Dawkins, "Animal Signals, Mind-reading and Manipulation," in J. Krebs and N. Davies, eds., *Behavioural Ecology*, 2nd ed. (Oxford: Blackwell, 1984). A. Whiten, ed., *Natural Theories of Mind* (Oxford: Blackwell, 1991). K. Sterelny, "Primate Worlds," in C. Heyes and L. Huber, eds., *The Evolution of Cognition* (Cambridge, MA: MIT, 2000). J. Call et al., "Domestic dogs (*Canis familiaris*) are sensitive to the attentional state of humans," *Journal of Comparative Psychology* 117, no. 3 (2003): 257–263.

13. Quoted in S. Baron-Cohen, *Mindblindness: An Essay on Autism and Theory of Mind* (Cambridge, MA: Bradford/MIT, 1995), 4–5. For reservations about this approach, see H. Tager-Flusberg, "What neurodevelopmental disorders can reveal about cognitive architecture," in P. Carruthers et al., eds., *The Innate Mind*.
14. A. Turing, "Computing machinery and intelligence," *Mind* 59 (1950): 433–460. In Turing's original paper "often enough" is cashed out as at least as often as a human interlocutor can fool interrogators that he or she is a member of the opposite sex.
15. The most discussed refutation, or attempted refutation, of Turing is John Searle's famous Chinese room thought experiment, which has spawned a large literature. The locus classicus for the argument is J. Searle, "Minds, brains, and programs," *Behavioral and Brain Sciences* 3 (1980): 417–424. This volume also contains a set of replies.
16. Cited in M. Cartmill, *A View to a Death in the Morning: Hunting and Nature Through History* (Cambridge, MA: Harvard, 1993), 40.
17. Cartmill, *A View to a Death in the Morning.*
18. G. Davies et al., eds., *Perceiving and Remembering Faces* (London: Academic Press, 1981). T. M. Field and N. A. Fox, eds., *Social Perception in Infancy* (Norwood, NJ: Ablex, 1985). E. H. Gombrich, *Art and Illusion: A Study in the Psychology of Pictorial Representation*, 2nd ed. (Princeton: Princeton University Press, 1969), 103. S. Platek et al., "Neural correlates of facial resemblance," *NeuroImage* 25 (2005): 1336–1344.
19. For primate vision and social interaction, see J. Allman, *Evolving Brains* (New York: Scientific American Library, 1999); G. C. Conroy, *Reconstructing Human Origins* (New York: Norton, 2005); and C. R. Badcock, "Mentalism and Mechanism: The twin modes of human cognition," in C. Crawford and C. Salmon,

eds., *Human Nature and Social Values: Implications of Evolutionary Psychology for Public Policy* (Mahwah, NJ: Lawrence Erlbaum, 2004). For the EDD, see Baron-Cohen, *Mindblindness,* and S. O'Connell, *Mindblindness: How We Learn to Love and Lie* (London: Heinemann, 1997).

20. This command is usually attributed to either William Prescott or Israel Putnam at the Battle of Bunker Hill. By the time of the battle, the expression had already been in circulation for over thirty years. It apparently was first used in 1743 by a Scotsman, Sir Andrew Agnew of Lochnaw, who told his men, "Dinna fire till ye can see the whites of their e' en . . . if ye dinna kill them they'll kill you."
21. D. A. Grossman, *On Killing: The Psychological Cost of Learning to Kill in War and Society* (New York: Little Brown, 1995), 128. For Miron and Goldstein's research, see M. S. Miron and A. P. Goldstein, *Hostage* (New York: Pergamon, 1979).
22. J. T. McCurdey, *War Neuroses* (London: Cambridge University Press, 1918), 129.
23. J. U. Nef, *War and Human Progress: An Essay on the Rise of Industrial Civilization* (New York: Norton, 1968).
24. C. Miles-Sheehan, *CNN Presents: Fit to Kill,* October 26, 2003.
25. "One hell of a big bang," *Guardian,* August 6, 2002.
26. J. Bourke, *An Intimate History of Killing* (New York: Basic Books, 1999). F. P. Crozier, *The Men I Killed* (London: M. Joseph, 1937), 101.
27. J. Hatzfeld, *Machete Season: The Killers in Rwanda Speak* (New York: Farrar, Straus & Giroux, 2005), 21–22. Compare with the following lines by a M. Grover, a veteran of the Boer war (from M. V. W. Smith, *Drummer Hodge: The Poetry of the Anglo-Boer War, 1899–1902* [Oxford: Oxford University Press, 1975], 101).

    I killed a man at Graspan,
    I killed him fair in fight
    And the Empire's poets and the empire's priests
    Swear blind I acted right. . . .
    But they can't stop the eyes of the man that I killed
    From starin' into mine.

28. Grossman, *On Killing,* 31.

10. PREDATORS, PREY, AND PARASITES

1. D. Hume, *A Treatise of Human Nature* (London: Penguin, 1985), 397.
2. Bob MacGowan, *CNN Presents: Fit to Kill,* October 26, 2003.

3. S. B. Hrdy, *Mother Nature: Maternal Instincts and How They Shape the Human Species* (New York: Ballantine, 1999), 465–466.
4. S. Keen, *Faces of the Enemy: Reflections of the Hostile Imagination* (New York: HarperCollins, 1986).
5. For examples, see S. Keen, *Faces of the Enemy*, 2nd ed. (San Francisco: HarperCollins, 1991).
6. S. Sondheim, *Sweeney Todd: The Demon Barber of Fleet Street* (New York: Applause Books, 2000).
7. D. Hart, "Humans as prey," *Chronicle of Higher Education*, April 21, 2006. For an excellent review of predation on humans, from prehistory to the present, see D. Hart and R. W. Sussman, *Man the Hunted: Primates, Predators and Human Evolution* (New York: Westview, 1905).
8. P. Shepherd, *The Others: How Animals Made Us Human* (Washington, DC: Island Press/Shearwater Books, 1996),29. D. Quammen, *Monster of God: The Man-Eating Predator in the Jungles of History and the Mind* (New York: Norton, 2003), 329.
9. B. Ehrenreich, *Blood Rites: The Origins and History of the Passions of War* (New York: Henry Holt, 1997).
10. J. A. Byers, *American Pronghorn: Social Adaptations and the Ghosts of Predators Past* (Chicago: University of Chicago Press, 1998). L. A. Isbell, "Snakes as agents of evolutionary change in primate brains," *Journal of Human Evolution* 51 (2006): 1–35. J. Blumstein et al., "How does the presence of predators influence the persistence of antipredator behavior?" *Journal of Theoretical Biology* 293 (2006): 460–468.
11. Ehrenreich, *Blood Rites*, 87.
12. H. C. Barrett, "Adaptations to predators and prey," in D. Buss, ed., *The Handbook of Evolutionary Psychology* (Hoboken, NJ: Wiley, 2005).
13. Ibid., 209–210.
14. G. Sajer, *The Forgotten Soldier* (Washington, D.C.: Brassey's, 2000).
15. Homer *Illiad*, trans. W. H. D. Rouse (Harmondsworth: Penguin, 1950), 201.
16. Ibid., 71.
17. Ibid., 200.
18. Ibid., 80.
19. Ibid., 194.
20. R. Kipling, *Morning Post*, June 22, 1915. T. Fleming, *The Illusion of Victory: America in World War I* (New York: Basic Books, 2003).
21. R. Fisk, "Talks with Osama Bin Laden," *Nation*, September 21, 1998.
22. R. Holmes, *Acts of War: The Behavior of Men in Battle* (London: Free Press, 1985).

23. A. Evans-Pritchard, "US asks Nato for help in 'draining the swamp' of global terrorism," *Telegraph*, September 21, 2001.
24. D. Winter, *Death's Men: Soldiers in the Great War* (London: Penguin, 1993), 91. E. Jünger, *The Storm of Steel: From the Diary of a German Storm-Troop Officer on the Western Front* (London: Penguin, 2004), 276.
25. J. Bourke, *An Intimate History of Killing* (New York: Basic Books, 1999), 50.
26. J. W. Dower, *War Without Mercy: Race and Power in the Pacific War* (New York: Pantheon, 1986), 90.
27. Ibid.
28. N. A. D. Armstrong, *Fieldcraft, Sniping and Intelligence*, 5th ed. (Aldershot: Gale & Polden, 1942), vii. J. Bourke, *An Intimate History of Killing*.
29. J. Bourke, *An Intimate History of Killing*, 222.
30. J. Glover, *Humanity: A Moral History of the 20th Century* (New Haven, CT: Yale University Press, 2000), 130.
31. A. Santoli, *Everything We Had: An Oral History of the Vietnam War by Thirty-Three American Soldiers Who Fought It* (New York: Ballantine, 1981), 98–99.
32. *Aberdeen Saturday Pioneer*, December 20, 1891, cited in D. E. Stannard, *American Holocaust: The Conquest of the New World* (Oxford: Oxford University Press, 1992).
33. A. Pagden, *The Fall of Natural Man* (Cambridge: Cambridge University Press, 1982), 23. G. Jahoda, *Images of Savages: Ancient Roots of Modern Prejudice in Western Culture* (London: Routledge, 1999), 21–46. F. Chalk and K. Jonassohn, *The History and Sociology of Genocide: Analyses and Case Studies* (New Haven, CT: Yale University Press, 1990). L. S. Dawidowicz, *The War Against the Jews, 1933–1945* (New York: Bantam, 1986). T. Segev, *Soldiers of Evil: The Commandants of the Nazi Concentration Camps* (New York: McGraw-Hill), 139. J. Hatzfeld, *Machete Season: The Killers in Rwanda Speak* (New York: Farrar, Straus & Giroux, 2005). Oddly, the Rwandan genocide often used botanical metaphors for killing such as "cutting down the tall trees," "bush clearing," and "uprooting weeds."
34. L. Keeley, *War Before Civilization: The Myth of the Peaceful Savage* (Oxford: Oxford University Press, 1996).
35. Ibid., 101.
36. I Samuel 18:27, I Samuel 17:51, 54; 31:9; II Samuel 20: 22.
37. V. A. Andrushko et al. "Bioarcheological evidence for trophy-taking in prehistoric central California," *American Journal of Physical Anthropology* 127 (2005): 375–384. Keely, *War Before Civilization*. Herodotus, *Histories* (London: Penguin, 2003).
38. S. A. LeBlanc, *Constant Battles: The Myth of the Peaceful, Noble Savage* (New York: St. Martin's Press, 2003).

39. Bourke, *An Intimate History of Killing.* M.P. Motley, *The Invisible Soldier: The Experience of Black Soldiers in World War II* (Detroit: Wayne State University Press, 1987).
40. Dower, *War Without Mercy.* E. L. Jones, "One war is enough," *Atlantic Monthly*, February 1946, 49.
41. Ibid.
42. Bourke, *An Intimate History of Killing.*
43. M. Baker, *Nam: The Vietnam War in the Words of the Men and Women Who Fought There* (London: Cooper Square Press, 1981), 50.
44. Cited in H. Zinn, A *People's History of the United States, 1492–Present* (New York: HarperCollins, 2001), 315.
45. C. B. Stanford, *The Hunting Apes: Meat Eating and the Origins of Human Behavior* (Princeton, NJ: Princeton University Press, 1999). J. Van Hooff, "Intergroup competition and conflict in animals and man." in J. Van der Dennen and V. Falger, eds., *Sociobiology and Conflict: Evolutionary Perspectives on Competition, Cooperation, Violence and Warfare* (London: Chapman and Hall, 1990). H. C. Barrett, "Adaptations to predators and prey."
46. Keeley, *War Before Civilization.* K. R. Otterbein, *How War Began* (College Station, TX: Texas A&M University Press, 2004). R. A. Marlar et al., "Biochemical evidence of cannibalism at a prehistoric Puebloan site in southwestern Colorado," *Nature* 407 (2000): 74–78.
47. Ehrenreich, *Blood Rites*, 10–11.
48. Ibid., 11.
49. Bourke, *An Intimate History of Killing*, 221.
50. Ibid.
51. Dower, *War Without Mercy*, 90.
52. "The reality of war," *Observer*, March 30, 2003.
53. Dower, *War Without Mercy*, 86.
54. I. Chang, "The rape of Nanking," in A. L. Barstow, ed., *War's Dirty Secret: Rape, Prostitution and Other Crimes Against Women* (Cleveland, OH: Pilgrim Press, 2000), 48.
55. Cited in D. A. Grossman, *On Killing: The Psychological Cost of Learning to Kill in War and Society* (New York: Little Brown, 1995), 236.
56. Holmes, *Acts of War*, 376. "Lyndon Johnson: 1908–1973," *Time*, February 5, 1973. T. E. Ricks, "War could last months, officers say," *Washington Post*, March 27, 2003. It is also interesting to note that both military organizations and well as combat units adopt the names of predators. For example, Zeljko Raznatovic's Serbian "Tigers," the "Tamil Tigers, Abul Waleed's Wolf Brigade militia, and U.S. Aviation squadrons such as the Black Lions, Blue Wolves,

Tigers, Wildcats, Red Lions, Wolfpack, Jaguars, Grey Wolves, and Cougars. Americans soar into battle in aircraft with names like Harrier, Hellcat, Cougar, Raptor, and Tigershark, while the Chinese pilot the Flying Leopard and Fierce Dragon.

57. A. Hitler, *Hitler's Table Talk* (Oxford: Oxford University Press, 1988), 332.
58. C. Wills, *Yellow Fever, Black Goddess: The Coevolution of People and Plagues* (Boston: Addison Wesley, 1997). P. E. Ewald, *Evolution of Infectious Disease* (Oxford: Oxford University Press, 1996). W. McNeill, *Plagues and Peoples* (New York: Anchor, 1998).
59. J. H. Park et al., "Evolved disease-avoidance processes and contemporary anti-social behavior: Prejudicial attitudes and avoidance of people which physical disabilities," *Journal of Non-Verbal Behavior* 27, no. 2 (2003): 65–87.
60. W. I. Miller, *The Anatomy of Disgust* (Cambridge, MA: Harvard University Press, 1997), 158–159.
61. P. Rozin and A. E. Fallon, "A perspective on disgust," *Psychological Review* 94 (1987): 23–41. P. Rozin, J. Haidt, and C. R. McCauley, "Disgust," in M. Lewis and J. M. Haviland, eds., *Handbook of Emotions* (New York: Guilford, 1993). P. Rozin, "Food is fundamental, fun, frightening, and far-reaching," *Social Research* 66 (1999) 9-30. A. Angyal, "Disgust and related aversions," *Journal of Abnormal and Social Psychology* 36 (1941): 393–412.
62. P. Rozin and A. E. Fallon, "A perspective on disgust," *Psychological Review* 94 (1987): 23–41. P. Rozin, J. Haidt, and C. R. McCauley, "Disgust," in Lewis and Haviland, eds., *Handbook of Emotions.* J. Haidt, C. R. McCauley, and P. Rozin, "Individual differences in sensitivity to disgust: A scale sampling seven domains of disgust elicitors," *Personality and Individual Differences* 16 (1994): 701–713. The Zoroastrian text is cited in P. Boyer, *Religion Explained: The Evolutionary Origins of Religious Thought* (New York: Perseus, 2001), 213.
63. Miller, *Anatomy of Disgust*, 137.
64. J. Diamond, *Guns, Germs and Steel: The Fates of Human Societies* (New York: Norton, 1997). V. Nutton, "Did the Greeks have a word for it?" in L. I. Conrad and D. Wujastyk, eds., *Contagion: Perspective from Pre-Modern Societies* (Burlington, VT: Ashgate, 2000). J. Falkner et al., "Evolved disease-avoidance mechanisms and contemporary xenophobic attitudes," *Group Processes and Intergroup Relations* 7, no. 40 (2004): 333—353.
65. D. C. Munro, "Urban and the Crusaders," *Translations and Reprints from the Original Sources of European History*, vol 1 (Philadelphia: University of Pennsylvania Press, 1895), 5–8.
66. Captain Wait Winthrop, cited in Chalk and Jonassohn, eds., *History and Sociology of Genocide*, 194.

67. P. Burrin, "Nazi antisemitism: Animalization and demonization," in R. S. Wistrich, ed., *Demonizing the Other: Antisemitism, Racism and Xenophobia* (Amsterdam: Harwood Academic Publishers, 1999), 226.
68. Dawidowicz, *War Against the Jews*, 19.
69. A. Hitler, *Mein Kampf*, trans. Ralph Manheim (London: Hutchinson, 1969), 53.
70. R. M. Lerner, *Final Solutions: Biology, Prejudice and Genocide* (University Park: Pennsylvania State University Press, 1992), 36.
71. *Der ewige Jude* (1940).
72. Dower, *War Without Mercy*, 91–92.
73. Ibid., 91.
74. B. Kiernan, "Twentieth-century genocides: Underlying ideological themes from Armenia to East Timor," in R. Gellately and B. Kiernan, eds. *The Specter of Genocide: Mass Murder in Historical Perspective* (Cambridge: Cambridge University Press, 2003), 33. E. Kissi, "Genocide in Cambodia and Ethiopia," in Gellately and Kiernan, B., eds. *Specter of Genocide*. Chalk Jonassohn, eds., *History and Sociology of Genocide*, 404.
75. L. Dwyer and D. Santikarma, "'When the world turned to chaos': 1965 and its aftermath in Bali, Indonesia," in R. Gellately and B. Kiernan, eds., *Specter of Genocide*.
76. B. Kiernan, "Twentieth-century genocides," in R. Gellately and B. Kiernan, eds., *Specter of Genocide*, 33.
77. J. Hatzfeld, *Machete Season*, 231.
78. J. Glover, *Humanity*, 130.
79. H. Bosmajian, "Dehumanizing people and euphemizing war," *Christian Century*, December 5, 1984.
80. R. Chandrasekaran, "Marines bide their time in insurgent-held Fallujah: Officers say Iraqi army must be fit to retake city," *Washington Post*, September 21, 2004.
81. J. Semelin, "Analysis of a mass crime: ethnic cleansing in the former Yugoslavia, 1991–1999," In R. Gellately and B. Kiernan, eds. *Specter of Genocide*.
82. "Post-war mopping up," *US News and World Report*, June 9, 2003.

11. HUMANITY LOST AND FOUND

1. For example, J. Keegan, *The Face of Battle: A Study of Agincourt, Waterloo, and the Somme* (Harmondsworth: Penguin, 1983), and R. L. O'Connell, *Ride of the Second Horseman: The Birth and Death of War* (Oxford: Oxford University Press, 1995).

2. J. G. Gray, *The Warriors: Reflections on Men in Battle* (Lincoln: University of Nebraska Press, 1970), 52
3. P. Caputo, A *Rumor of War* (New York: Henry Holt, 1977), 254; cited in J. Bourke, *An Intimate History of Killing* (New York: Basic Books, 1999), 340. C. Hedges, *War Is a Force That Gives Us Meaning* (New York: Anchor, 2002), 100.
4. R. Chisholm, *Cover of Darkness* (London: Elmfield Press, 1976), 71. M. Pottle, ed., *Champion Redoubtable: The Diaries and Letters of Violet Bonham Carter, 1914–1945* (Weidenfeld & Nicolson: London, 1998), 25.

APPENDIX

1. S. Totten, W. S. Parsons, and R. K. Hitchcock, "Confronting genocide and ethnocide of indigenous peoples: An interdisciplinary approach to definition, intervention, prevention and advocacy," in A. L. Hinton, ed., *Annihilating Difference: The Anthropology of Genocide* (Berkeley, CA: University of California Press, 2002).

INDEX